NEWTONIAN
ELECTRODYNAMICS

NEWTONIAN ELECTRODYNAMICS

Peter Graneau

Northeastern Univ. Boston

Neal Graneau

Oxford Univ.

World Scientific
Singapore • New Jersey • London • Hong Kong

Published by

World Scientific Publishing Co. Pte. Ltd.

5 Toh Tuck Link, Singapore 596224

USA office: 27 Warren Street, Suite 401-402, Hackensack, NJ 07601

UK office: 57 Shelton Street, Covent Garden, London WC2H 9HE

British Library Cataloguing-in-Publication Data
A catalogue record for this book is available from the British Library.

NEWTONIAN ELECTRODYNAMICS

ISBN 978-981-02-2284-0 (hardcover)
ISBN 978-981-02-2681-7 (paperback)
ISBN 978-981-283-093-7 (ebook for institutions)
ISBN 978-981-310-312-2 (ebook for individuals)

For any available supplementary material, please visit
https://www.worldscientific.com/worldscibooks/10.1142/2770#t=suppl

To Brigitte Graneau,

without whose love and support

this book would not exist

CONTENTS

PREFACE

The Newtonian electrodynamics, due primarily to Ampère, F.E. Neumann and Kirchoff, was of critical importance for the evolution of the electrical age. In the twentieth century, the old theory with modern extensions continues to be productive in the field of electro-mechanics and arc dynamics. It provides the foundations of what electrical engineers call 'circuit theory'. Electro-mechanics and certain aspects of circuit theory cannot be grasped and analyzed without Newtonian concepts.

Teachers have the noble desire of treating all of electromagnetism with the relativistic field theory of Maxwell, Lorentz and Einstein. From a pedagogic point of view, the urge to unify physics is more than understandable. In order to achieve it, the Newtonian electrodynamics had to be swept under the carpet and written out of textbooks. Not foreseen by field theoreticians were the gaping holes that this left in the comprehension of such basic elements of electrical engineering as motors and generators. We question the wisdom of driving unification so far that areas of an important science are left without the quantitative basis for doing applied physics and designing machines.

It is worth remembering that the whole world -- space travellers included -- relies on Newton's principles and laws of mechanics, even though they are not supposed to represent the true workings of nature. Instead Einstein's general theory of relativity is currently considered to describe reality. Nevertheless, pragmatic scientists and engineers have not seen the need to abandon the Newtonian theory which is simpler to use and works so well. After reading our book, pragmatic electrodynamicists will reach the same conclusion. In the twenty-first century, students of electromagnetism may have no choice but to become familiar with Newtonian electrodynamics as well as field theory.

The last two chapters on arc dynamics break new ground. There are hardly any books available in this field. The reason undoubtedly is the failure of the Lorentz force to correctly describe arc forces. The exciting research on the liberation of internal water energy, described in Chapter 7, is still continuing, but we are only able to relate the findings up to October 1st, 1995, in order to make the book available for printing.

Of the scientists and engineers who have contributed to the revival of the Newtonian electrodynamics, and who are recognized in the many references, we would like to give special thanks to Dr. Thomas E. Phipps, Jr., for his theoretical and experimental contributions. Unfortunately, his paper "A do-it-yourself refutation of modern physics" came too late to be included in the book. We would also like to thank Julie and Alex Hammond for their painstaking proofreading of the manuscript.

Concord, Massachusetts   Oxford, England

October 1995

Newton's First Rule of Reasoning:

"We are to admit no more causes of natural things than such as are both true and sufficient to explain their appearances"

Evolution of the Nineteenth Century Newtonian Electrodynamics

The Birth of Electromagnetism

The concepts of electricity and magnetism have existed since the time of the ancient Greeks. Since then it has been believed that the attraction of iron by lodestone and of many kinds of matter by electrified amber had something in common. At least since the Middle Ages man had known that when lightning struck iron it could imbue this metal with magnetism. By the same token, the fire from heaven was capable of changing the polarity of a compass needle. Dibner [1.1] reported that in 1802, Romagnosi, a lawyer and physicist at the University of Parma in Italy, reversed the polarity of a compass by passing a galvanic current along the needle. This experiment came close to the discovery of electromagnetism which has been universally attributed to the Danish scientist Hans Christian Oersted (1777-1851), eighteen years later.

Oersted, a professor of natural philosophy in Copenhagen, determined the direction in which a compass needle would turn when a straight wire with electric current flowing along it was brought near to the needle without touching it. One might ask why this particular experiment was singled out as the beginning of electromagnetism?

Oersted felt so certain of the enthusiastic reception of his discovery that he had a paper printed for the occasion and sent to all scientists and journals of note [1.2] The paper was dated July 21, 1820. It claimed that 'magnetic flux' encircled the current, but Oersted called this flux 'electric conflict'. Here was the missing link between electricity and magnetism

It has to be remembered that Oersted's explanation of the magnetic influence of an electric current came at a time when effluvia, ether, and ether vortices were not in vogue because of the success of Newton's and Coulomb's action at a distance laws which avoided any reference to what was happening in the space between interacting bodies. Newton's own words serve best to describe the prevailing philosophy of contemporary natural philosophers. In the preface to the first edition of the 'Principia' he said:

> "Then from these forces, by other propositions which are also mathematical, I deduce the motion of the planets, the comets, the moon and the sea. I wish we could derive the rest of the phenomena of Nature by the

same kind of reasoning from mechanical principles, for I am induced for many reasons to suspect that they may all depend upon certain forces by which the particles of bodies, by some causes hitherto unknown, are either mutually impelled towards one another, and cohere in regular figures, or are repelled and recede from one another. These forces being unknown, philosophers have hitherto attempted the search of Nature in vain; but I hope the principles here laid down will afford some light either to this or some truer method of philosophy."

These words apply equally to the Ampère-Neumann electrodynamics which followed on the heels of Oersted's experiment. Neither Ampère nor F.E. Neumann took heed of the magnetic circles around electric currents. They wrote down the laws of electrodynamic force according to the far-action model of Newtonian gravitation. Newton was more precise and spoke of mutual simultaneous interaction between particles so as not to create the impression of something travelling at finite or infinite speed between the interacting elements of matter.

In 1820, as today, there lived many scientists who disliked action at a distance. In their hearts they had adhered to the Aristotelian principle that matter cannot act where it is not. Oersted felt many of them would welcome an electromagnetic theory in terms of an active field and field-contact action. Forty years later Maxwell placed this electromagnetic field on mathematical foundations, and within a few decades the world of physics had shed all of its remote action concepts.

Oersted's announcement [1.2] triggered a frenzy of activity in Paris which had been established by Napoleon as the world's capital of science. It caused the French Academy to stage a demonstration of the Copenhagen experiment. As Hammond [1.3] recounts the event, Ampère was present and went straight home after the demonstration to begin work on a new science which he called 'electrodynamics'. He even left before the discussion at the Academy, at which he was a regular contributor. This was September 11, 1820. Precisely one week later, Ampère read a paper before the Academy and reported that parallel wires carrying electric currents attract or repel each other, depending on whether the two currents flow in the same or opposite directions. This was as great a leap forward in electromagnetism as Oersted's. Ampère followed up with weekly presentations to the Members of the Academy of the progress he was making in his experimental investigation of the interaction of electric currents. In only a few months he had laid the foundations of the new science of electrodynamics.

Like Ampère, Jean-Baptiste Biot (1774-1862) was also a professor in Paris. He was an expert in the measurement of the strength of the earth's magnetic field. The frequency of oscillation of a compass needle had been found to be a measure of the field strength. Biot had accompanied Gay-Lussac on the first balloon flight in order to determine if the earth's magnetic field varied with height above ground level. Biot was another French scientist who was present at the Academy meeting of September 11 and, like Ampère, he too rushed back to his laboratory. With his assistant Felix Savart he set up a galvanic current in a long vertical wire. With due compensation for the terrestrial field, the two proceeded to survey the magnetic field strength around the wire using their well established method. Figure 1.1(a) illustrates the nature of the Biot-Savart result. The force H which would be exerted on a unit magnetic pole was found to be inversely proportional to the shortest distance r to the wire. These investigators had no means of measuring the strength of the current i, as the galvanometer was still to be invented by Ampère. It is not clear, therefore, if the proportionality of H to i was taken for granted or established by a later series of measurements. At any rate, the Biot-Savart experiments led to the following well-known formula for straight conductors

$$H = k \frac{i}{r} \tag{1.1}$$

where k is a dimensional constant. The units of the early electrodynamics were the fundamental electromagnetic units based on the centimeter, the gram and the second (e.m.u).

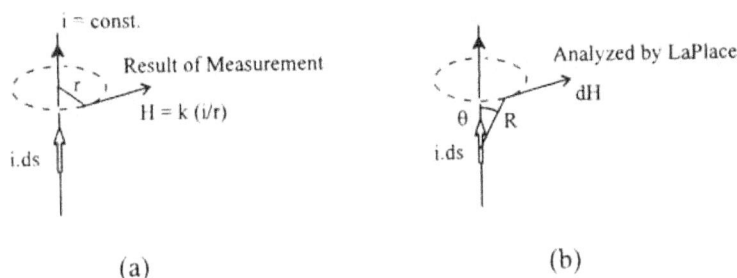

(a)   (b)

Figure 1.1 : The Biot-Savart Law

Biot and Savart spoke of their findings to the French Academy on October 30, 1820. They were obviously competing with Ampère in the unravelling of further electromagnetic mysteries. What today is understood to be the Biot-Savart law is not Eq.1.1, but the differential of it with respect to the current element i.ds, as shown in figure 1.1(b). The law may be written

$$dH = \frac{k}{R^2} i \, ds \, \sin \theta \tag{1.2}$$

where k is again a dimensional constant. According to Biot and Ampère, Eq.1.2 was derived from Eq.1.1 by LaPlace, who never claimed credit for it. An excellent account of these happenings has been written by Tricker [1.4, 1.5].

The Biot-Savart law introduces the concept of the 'current-element' which has become the 'particle' of the Ampère-Neumann electrodynamics. Without subdividing the wire into small elements it would have been impossible to compute the magnetic field strength at a point due to a closed circuit. Ampère employed the same current element in his force law. Before considering Ampère's work, a little has to be said about the status of electrostatics at the beginning of the nineteenth century.

Whittaker in his 'History of the theories of aether and electricity' [1.6] maintains:

> "By Franklin's law of the conservation of electric charge, and Priestley's law of attraction between charged bodies, electricity was raised to the position of an exact science."

Benjamin Franklin working in isolation in America propounded the one-fluid theory of electricity which may be taken as the natural precursor of our present electronic theory of solids. He also came to the conclusions that certain substances, which we now describe as dielectrics, are impenetrable to electric effluvia. Therefore the two electrodes of a Leyden jar had to communicate by far-action. Through his friend Priestley in England, Franklin's work was published in Europe. Particularly his experiments of drawing lightning out of thunderclouds received much attention. During the 1750s the German physicist Aepinus took up Franklin's ideas and with them elucidated the phenomenon of electrostatic induction.

In 1766 Franklin wrote to Priestley asking him to repeat an experiment with cork balls in an electrically charged metal container. Franklin had found to his surprise that the cork balls did not respond to the charge. Subsequently Priestley established that the electric field strength, as it would be called today, was zero inside an electrically charged closed metallic vessel. He clearly recognized the analogy to gravitation and concluded [1.6]:

> "May we not infer from this experiment that the attraction of electricity is subject to the same laws with that of gravitation, and is therefore according to the square of the distances; since it is easily demonstrated that were the earth in the form of a shell, a body in the inside of it would not be attracted to one side more than to another?"

The inverse-square-law of the interaction of electric charges did, however, not become common property of the scientific community of the eighteenth century until Coulomb proved it directly in 1785 by measurements with a torsion balance. Coulomb's law may be written

$$F = k \frac{q_1 q_2}{r^2}$$

(1.3)

The charges q_1 and q_2 are separated by the distance r between their centers and k is a dimensional constant. If the charges are both positive or both negative, the force F is positive and this represents repulsion. For charges of unlike sign the force is negative which stands for attraction.

Ampère's Force Law

In the tradition of the great French mathematicians, who developed the science of mechanics from Newton's laws, Ampère set out to cast electromagnetism in a Newtonian mould. For this he required an appropriate fundamental law of the interaction of electrodynamic matter elements. He suspected this would turn out to be an inverse-square-law akin to those of Newton and Coulomb which, to use his own words in English translation [1.4]: ".... opened a new highway into the sciences which have natural phenomena as their object of study."

However difficult it may appear today, the question of what constituted the elementary 'particle' of electrodynamics apparently posed no problem to Ampère, Biot and Savart. They all employed the metallic current-element. It is uncertain who may have thought of this concept first. Ampère clearly recognized that, unlike the elementary particles of gravitation and electrostatics, which were characterized by a simple scalar magnitude (of mass or charge), the current-element would in addition to its magnitude of current strength have to possess length and direction.

On the basis of his first electrodynamic experiments, showing the attraction and repulsion of straight and parallel current carrying wires, Ampère expected the law of mechanical force between two current elements to be of the general form

$$\Delta F_{m,n} = - i_m i_n \frac{dm \; dn}{r_{m,n}^2} f(\alpha, \beta, \varepsilon) \qquad (1.4)$$

The Δ in Eq.1.4 infers that we are dealing with an elemental force which cannot be measured directly because current elements of wires are not available in isolation. The forces that are measured in the laboratory are sums of many elemental forces. In Eq.1.4 the elements carry currents of i_m and i_n, and their lengths are dm and dn. The distance between the center points of the elements is $r_{m,n}$, and the angles of the function f are shown in figure 1.3.

If the angle function $f(\alpha, \beta, \varepsilon)$ is positive, then $\Delta F_{m,n}$ is negative, which indicates attraction between the elements. Ampère originally proposed the opposite sign convention, but this was subsequently dropped in order to coordinate Eq.1.4 with Coulomb's law, Eq.1.3. Both the current strengths and element lengths were taken to be positive scalar quantities, while the directional properties of the elements were given by the angle function f.

With respect to the proportionality of the elemental force to the lengths and currents of the two elements Ampère said [1.4]:

"First of all, it is evident that the mutual action of two elements of electric current is proportional to their lengths; for assuming them to be divided into infinitesimal equal parts along their length, all attractions and repulsions of these parts can be regarded as directed along one and the same straight line, so that they necessarily add up. This action must also be proportional to the intensities of the two currents."

In his early papers on electrodynamics Ampère also assumed the proportionality of the elemental force to the inverse square of the distance of separation, because he believed all fundamental forces of nature concurred with this distance dependence. Later he proved the validity of this early assumption with the three-circle experiment of figure 1.2. For the sake of clarity the figure does not show the current leads to the three parallel and coaxial current circles in vertical planes. All three circles were connected in series to ensure equal current intensities in all of them. Ampère's method of compensating for the effect of the earth's magnetic field has also been omitted in figure 1.2. The radii of, and the distances between the three current circles were chosen such that the geometrical relationship of circle 1 to circle 2 was similar to the relationship of circle 2 to circle 3. In other words, the only difference between the 1-2 and 2-3 combinations was a linear scale factor. Circles 1 and 3 were fixed to the laboratory frame, while circle 2 was held coaxial with the other two, but with its insulator arm free to rotate about the vertical line YY. The purpose of this experiment was to show that, if the currents in 1 and 3 encircled the common axis in the same direction, circle 2 would remain stationary, for it was either attracted or repelled equally strongly by the two adjacent circles. When the current in circle 2 flowed in the direction shown on figure 1.2, the force was repulsion.

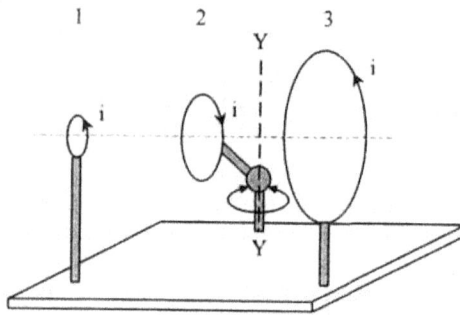

Figure 1.2 : Ampère's three-circle experiment

    This experiment proved that the reciprocal electrodynamic forces between two current loops were independent of the linear scale factor and therefore independent of the size of the electrodynamic system of conductors. The same had to be true for all elemental forces into which the total force could be divided. Hence the geometrical factor $dm \ dn / r_{m,n}^2$ of Eq. 1.4 had to be a dimensionless number, a condition which could only be fulfilled by the inverse square law. In addition to this proof by the three-circle experiment, the widely known fact that geometrically similar small and large conductor arrangements are subject to the same mechanical forces for the same currents, further corroborates Ampère's conclusion.

    Not surprisingly, Ampère's most challenging task proved to be the determination of the angle function $f(\alpha, \beta, \varepsilon)$. A recurrent difficulty for those who have tried to understand Ampère's force law, and others who have used it for engineering calculations, has been the visualization of the three angles, particularly when the two elements do not lie in the same plane. Expressing the law in vector form does not eliminate the problem. Figure 1.3 attempts to make the visualization as easy as possible. M and N are the center points of two unequal current elements. The distance between M and N, that is $r_{m,n}$, must be treated as a vector. The polarity of this vector is arbitrary. It may be chosen to point from M to N, or from N to M. The current elements must also be treated as vectors and have to point in the direction of current flow. The angle through which the element $i_m dm$ has to be turned about M to make it point in the same direction as $r_{m,n}$ is α. Similarly, the angle through which the element $i_n dn$ has to be turned to point in the same direction as $r_{m,n}$ is β. Since both of these angles appear in cosines, and since $\cos\alpha = \cos(2\pi - \alpha)$, it does not matter in which direction the element vectors are turned to make them coincide with the distance vector. Each element and the distance vector lie in a plane of their own. The two planes intersect along the distance vector. Of the two complimentary angles between the planes, γ is that angle through which the plane containing $i_m \, dm$ would have to be turned in the direction indicated, in order to make the components of the current elements, which are perpendicular to the distance vector, point in the same direction.

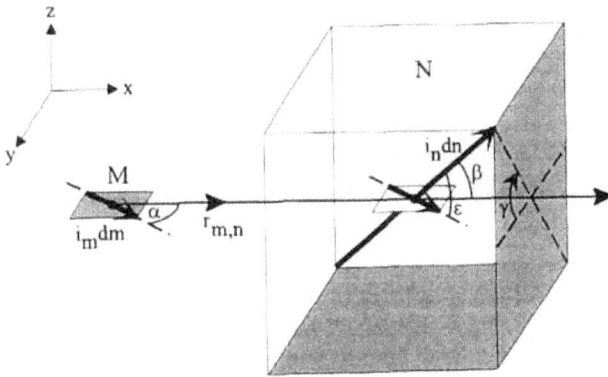

Figure 1.3 : Angles in Ampère's force formula

Another important angle in Ampère's formula is ε. It stands for the angle of inclination of the two current elements toward each other. It may best be visualized by transferring one of the elements parallel to itself along MN until its center coincides with the center of the other element. In figure 1.3 the dm element has been transferred from M to N and ε is the angle through which the transferred element has to be turned about N to make it point in the same direction as dn. Since ε also appears in a cosine, its direction of rotation is as arbitrary as that of α and β.

To see how Ampère determined $f(\alpha, \beta, \varepsilon)$ we resolve the two current elements of figure 1.3 into their cartesian components shown in figure 1.4. The elements $i_m \, dm$ and $i_n \, dn$ are there represented as vectors m and n, pivoted at the centers of the elements. The resolved components of the two current elements along the x, y, and z axes are given by

$$m(x) = i_m dm \, \cos \alpha \quad ; \quad m(y) = i_m dm \, \sin \alpha$$
$$n(x) = i_n dn \, \cos \beta \quad ; \quad n(y) = i_n dn \, \sin \beta \, \cos \gamma \quad ;$$
$$n(z) = i_n dn \, \sin \beta \, \sin \gamma$$

$$(1.5)$$

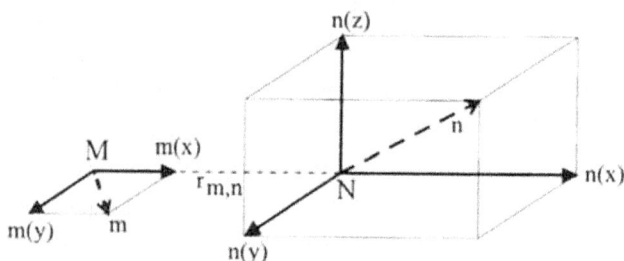

Figure 1.4 : Resolved component vector representation of the two general
current elements of figure 1.3

Now each component of m interacts with each component of n, resulting in a total of six contributions to the elemental force between two current elements. Four of them are zero according to a theorem first enunciated by Ampère. Generations of physicists have been uncertain how this theorem follows from Ampère's experiments or how it could be deduced from his postulates. We will refer to it as 'Ampère's Rule', leaving the question open of whether it is a theorem or an assumption. Ampère [1.7] wrote about it as follows:

> "An infinitely small portion of current exerts no action on another infinitely small portion of a current which is situated in a plane which passes through the midpoint and which is perpendicular to its direction. In fact, the two halves of the first element produce equal actions on the second, the one attractive and the other repellent, because the current tends to approach the common perpendicular in one of those halves and to move away from it in the other. These two equal forces form an angle which tends to two right angles according as the element tends to zero. Their resultant is therefore infinitesimal in relation to these forces and in consequence it can be neglected in the calculations."

In compliance with Ampère's Rule, the four vanishing force contributions of the element components drawn in figure 1.4 are

$$\Delta F_{m(x),n(y)} = \Delta F_{m(x),n(z)} = \Delta F_{m(y),n(x)} = \Delta F_{m(y),n(z)} = 0 \qquad (1.6)$$

A corollary of Ampère's Rule is that the mechanical interaction of two current elements arises from two sets of parallel element components, one of them being the set which lies along the line connecting the two elements, and the other is the set which is perpendicular to that line. Ampère then assumed that the two non-vanishing force contributions may be expressed by

$$\Delta F_{m(y),n(y)} = - \frac{m(y) \cdot n(y)}{r_{m,n}^2} \qquad (1.7)$$

$$\Delta F_{m(x),n(x)} = - k \frac{m(x) \cdot n(x)}{r_{m,n}^2} \qquad (1.8)$$

where k is a constant and the element components are defined by Eq.1.5. At this stage Eq.1.4 may be expressed as

$$\Delta F_{m,n} = - i_m i_n \frac{dm \ dn}{r_{m,n}^2} (\sin\alpha \ \sin\beta \ \cos\gamma + k \ \cos\alpha \ \cos\beta) \qquad (1.9)$$

Ampère then introduced the trigonometrical equation

$$\cos\varepsilon = \cos\alpha \ \cos\beta + \sin\alpha \ \sin\beta \ \cos\gamma \qquad (1.10)$$

For proof of Eq.1.10 he referred to a spherical triangle, but it may also be derived with the help of figure 1.5 from the direction cosines of the two general current elements. It is known that the cosine of the angle of inclination between two vectors is equal to the sum of the three products of corresponding direction cosines of the two vectors. With regard to figure 1.5 the direction cosines along the x, y and z axes of the dm-element are

$$\cos\alpha \quad , \quad \cos\left(\frac{\pi}{2} - \alpha\right) = \sin\alpha \quad , \quad \cos\delta = \cos\left(\frac{\pi}{2}\right) = 0$$

and those of the dn-element are

$$\cos \beta \;,\; \cos\left(\frac{\pi}{2} - \beta\right) \cos \gamma \;=\; \sin \beta \cos \gamma \;;\; \cos\left(\frac{\pi}{2} - \beta\right) \cos\left(\frac{\pi}{2} - \gamma\right) \;=\; \sin \beta \sin \gamma$$

Hence

$$\cos \varepsilon \;=\; \cos \alpha \; \cos \beta \;+\; \sin \alpha \; \sin \beta \; \cos \gamma \;+\; 0 \times \sin \beta \; \sin \gamma$$

which confirms Eq.1.10.

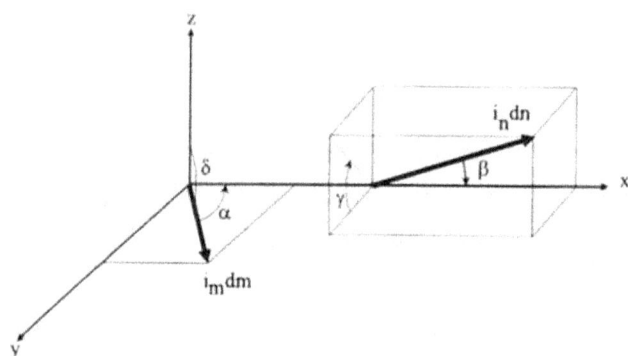

Figure 1.5 : Angles for determining the direction cosines of two general current elements

Attempting to deduce a force law for two co-planar elements, he used Eq.1.10, and the force formula Eq.1.9, to arrive at

$$\Delta F_{m,n} \;=\; -\; i_m \; i_n \frac{dm \; dn}{r_{m,n}^2} \; (\cos \varepsilon \;+\; (k-1) \; \cos \alpha \; \cos \beta) \tag{1.11}$$

After this step Ampère converted the cosines to partial differentials of $r_{m,n}$ with respect to small displacements of the centers of the elements, M and N, along the line of action. These partial differentials are further defined by figure 1.6. In the limit, as the displacements of M and N tend to zero, and writing r for the distance between the elements, we find that

$$\cos \alpha \;=\; \frac{\partial r}{\partial m} \;,\; \cos \beta \;=\; -\; \frac{\partial r}{\partial n} \tag{1.12}$$

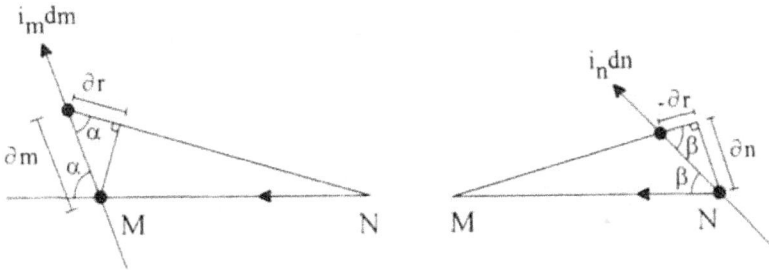

Figure 1.6. Partial differentials of the distance vector with respect to displacements of current elements

Furthermore, if M and N have the coordinates x_m, y_m, z_m, and x_n, y_n, z_n, we have

$$r^2 = (x_m - x_n)^2 + (y_m - y_n)^2 + (z_m - z_n)^2 \qquad (1.13)$$

Differentiating Eq.1.13 with respect to m results in

$$r \frac{\partial r}{\partial m} = (x_m - x_n) \frac{\partial x_m}{\partial m} + (y_m - y_n) \frac{\partial y_m}{\partial m} + (z_m - z_n) \frac{\partial z_m}{\partial m} \qquad (1.14)$$

and a second differentiation with respect to n gives

$$r \frac{\partial^2 r}{\partial m \, \partial n} + \frac{\partial r}{\partial m} \frac{\partial r}{\partial n} = - \frac{\partial x_m}{\partial m} \frac{\partial x_n}{\partial n} - \frac{\partial y_m}{\partial m} \frac{\partial y_n}{\partial n} - \frac{\partial z_m}{\partial m} \frac{\partial z_n}{\partial n} \qquad (1.15)$$

The right-hand side of Eq.1.15 contains the negative products of the direction cosines of the two current elements. Therefore

$$\cos \varepsilon = - r \frac{\partial^2 r}{\partial m \, \partial n} - \frac{\partial r}{\partial m} \frac{\partial r}{\partial n} \qquad (1.16)$$

Substituting Eq.1.12 and Eq.1.16 into the force equation, Eq.1.11, yields

$$\Delta F_{m,n} = i_m i_n \frac{dm\ dn}{r^2} \left(r \frac{\partial^2 r}{\partial m\ \partial n} + k \frac{\partial r}{\partial m} \frac{\partial r}{\partial n} \right) \tag{1.17}$$

This may also be written

$$\Delta F_{m,n} = i_m i_n \frac{dm\ dn}{r^2} \frac{1}{r^{k-1}} \frac{\partial}{\partial n} \left(r^k \frac{\partial r}{\partial m} \right)$$

$$= i_m i_n r^{(k-1)} \frac{\partial}{\partial n} \left(r^k \frac{\partial r}{\partial m} \right) dm\ dn \tag{1.18}$$

Ampère then invoked the result of another of his null-experiments to determine the value of k. The experiment to which he referred is sketched in figure 1.7. To distinguish it from the other null-experiments it will be called the wire-arc experiment. It proved that the mechanical force on a circular arc section of a current carrying circuit 1, due to current in a separate closed circuit 2 of any shape and disposition, was entirely perpendicular to the arc. As shown in figure 1.7, Ampère floated the arc section on two mercury troughs and left it free to rotate the insulator arm OX about the pivot O. During the experiment the arc remained stationary as circuit 2 was brought up to it and moved around. From this behaviour Ampère concluded that the net tangential force on the wire arc portion was zero.

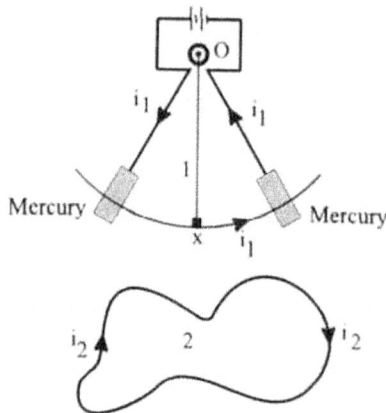

Figure 1.7 : Ampère's wire-arc experiment

Take Eq.1.18 and substitute for $\partial r/\partial m$ from Eq.1.12, leaving

$$\Delta F_{m,n} = i_m\, i_n\, dm\; r^{-(k+1)}\; \frac{\partial}{\partial n}\; (r^k\, \cos\alpha)\; dn \tag{1.19}$$

The component of the mutual force which acts tangentially on the dm-element is obtained by multiplying Eq.1.19 by $\cos\alpha$. If it is to agree with experiment, this tangential force, when integrated over all dn-elements in circuit 2, must come to zero. Hence we may write

$$\oint_2 \Delta F_{m,n}\, \cos\alpha = i_m\, i_n\, dm\; \oint_2 r^{-(2k+1)}\; r^k\, \cos\alpha\; \frac{\partial}{\partial n}\; (r^k\, \cos\alpha)\; dn = 0 \tag{1.20}$$

For integration by parts according to $\int u\, dv = uv - \int v\, du$, we let

$$u = r^{-(2k+1)} \quad ; \quad \frac{\partial u}{\partial n} = -(2k+1)\, r^{-2(k+1)}\, \frac{\partial r}{\partial n}$$

$$v = \frac{1}{2}\, r^{2k}\, \cos^2\alpha \quad , \quad dv = r^k\, \cos\alpha\; \frac{\partial}{\partial n}\; (r^k\, \cos\alpha)\; dn$$

Therefore

$$\oint_2 \Delta F_{m,n}\, \cos\alpha = \frac{1}{2}\, i_m\, i_n\, dm\; \left[\left. \left(\frac{\cos^2\alpha}{r} \right) \right|_{n'}^{n} + (2k+1) \oint_2 \frac{\cos^2\alpha}{r^2}\, dr \right] = 0 \tag{1.21}$$

The limits n and n' of the first term are actually adjacent infinitely short elements on circuit 2, and therefore the first term of Eq.1.21 vanishes. As Ampère pointed out, however, many closed circuits can be imagined for which the integral in the second term of Eq.1.21 will not vanish. Hence we are left with

$$k = -\frac{1}{2} \tag{1.22}$$

as the only possibility of reducing Eq.1.20 to zero, whatever the shape or disposition of circuit 2. As can be seen from Eq.1.7 and Eq.1.8, k determines the difference in the mutual interaction of equal parallel current elements between (a) elements lying along the line connecting their centers, and (b) elements set perpendicular to that line. With Eq.1.22 substituted into Eq.1.8, it is evident that two elements of unit strength and separated by unit distance repel each other half as strongly when lying along the distance vector, than they would attract each other when arranged transverse to this vector.

Using Eq.1.11 and Eq.1.22 Ampère wrote his force law as follows

$$\Delta F_{m,n} = - i_m \, i_n \, \frac{dm \, dn}{r_{m,n}^2} \, (\cos \varepsilon - \frac{3}{2} \cos \alpha \, \cos \beta) \tag{1.23}$$

It was Ampère who first clarified what was meant by voltage and current, and his force law indicated that the square of current must have the same dimension as mechanical force. With this knowledge he defined an electrodynamic unit of current which was smaller than the electromagnetic unit of current called the absolute-ampere. (1 ab-amp = 10 amps). To obtain these units of current, the electrodynamic measures have to be multiplied by $\sqrt{2}$. After this change of units we arrive at Ampère's force law for co-planar current elements, in its modern form

$$\Delta F_{m,n} = - i_m \, i_n \, \frac{dm \, dn}{r_{m,n}^2} \, (2 \cos \varepsilon - 3 \cos \alpha \, \cos \beta) \tag{1.24}$$

This gives the elemental force in dynes provided the currents i_m and i_n are inserted in absolute-amperes. If Eq.1.24 is further multiplied by $(\mu_0/4\pi)$, then the force is in Newtons when the current is in Amps.

A most important aspect of Ampère's theory is that the individual current element does not interact with itself. There is little or no discussion of this point in Ampère's papers because he took it for granted that in Newtonian science every elemental force is a mutual interaction of two elements of matter. Ampère considered his elements to be particles of the conductor metal and not travelling charges. This was in harmony with the ideas of the 1820s when electricity was still considered to be a subtile fluid or continuum. The conductor metal was considered to be infinitely divisible, and forces could therefore be determined with differential and integral calculus. Today the Ampère electrodynamics is being applied to the ions of the metal lattice or plasma, which imposes a lower limit on the size of the current element. Finite-size current elements are easily handled with computer assisted finite element analysis.

A common error made in modern treatments of Ampère's law is to assume that it represents the force between moving conduction electrons. One of the foremost scholars of Ampère's work was Maxwell. He stressed the fact that the Ampèrian current element is a stationary piece of metal and said [1.8]:

> "It must be carefully remembered, that the mechanical force which urges a conductor carrying a current across the lines of magnetic force, acts, not on the electric current, but on the conductor which carries it. .... The only force which acts on electric currents is electromotive force, which must be

distinguished from the mechanical force."

An often voiced contemporary criticism of current elements is that they produce discontinuities at straight element junctions. This fact was also of concern to Ampère. He went to great length to demonstrate to his own satisfaction that a smooth wire curve could be adequately represented with his discontinuous elements. One of his null-experiments was specifically designed to prove this argument. The essential features of the experiment are shown in figure 1.8. It will be referred to as the bent-wire experiment. AA'DE is a rectangular current loop in a vertical plane and suspended so that it is free to rotate about its vertical center line. As before, for the sake of clarity, Ampère's additional circuit to offset the terrestrial magnetic field is not shown in figure 1.8.

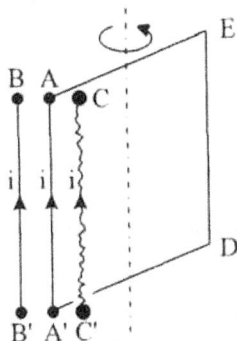

Figure 1.8 Ampère's bent-wire experiment

BB' is a straight wire parallel to AA' and placed close to it. CC' is the bent wire arranged with its axis parallel to and at the same level as BB' and AA'. Ampère fitted the bent wire into a narrow slot of a wooden post and described it as being twisted over its entire length in a plane perpendicular to BCC'B' and such that the wire at no point departed more than a very short distance from the center line of the slot. The experiment proved that if the wires BB' and CC' carried currents in the same direction and were equidistant from AA', as indicated in figure 1.8, no turning moment was exerted on the loop AA'DE. Hence, depending on the direction of the current in AA', the bent wire CC' exerted the same force of attraction or repulsion on the AA' wire as did the straight wire BB'.

Based on this experimental result, Ampère argued that a curved wire section, as for example AA' of figure 1.9, was equivalent to the straight section BB' provided B coincided with A and B' with A'. For an explanation he offered the vectorial cancellation of the transverse sub-elements. Ampère believed the argument also applied to a three-dimensional curve.

The published record for the few years in which Ampère concerned himself with electrodynamics shows little discussion of the current distribution over the wire cross-section and to what extent this may have been compatible with the single filament representation of a conductor. At that time conductors were usually thin wires, and the three-dimensional nature of the current stream was not a pressing issue. As will be shown later, Ampère's theory can be adapted to large conductors by subdividing them into current elements of finite volume and filaments of finite cross-section.

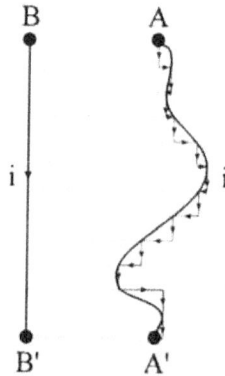

Figure 1.9 : Equivalence of bent and straight wire sections

Energy conservation was only introduced after Ampère's death, however electrodynamic forces are capable of doing work and they must therefore be associated with stored energy. Following Ampère's successes, this issue was taken up by F.E. Neumann (1798-1895).

Eq.1.24 is Ampère's empirical force law which he formulated to agree with the large body of his experimentation [1.7]. The law is not known to have failed during more than 170 years of its existence, so long as it is applied to metallic conductors for which it was created. The method by which Ampère deduced his law from experiments is only of academic interest. It has no bearing on the validity of the law. Like the scaffolding used when erecting a building, the method of deducing an empirical law may be discarded as soon as the law has been found.

Neumann's Electrodynamic Potential

Twenty years elapsed between the conclusion of Ampère's study of electrodynamics and F.E. Neumann's first memoir, in 1845, giving the original theory of electromagnetic induction. The author, Franz Neumann was the father of Carl Neumann, who also became a famous 'electrician' of the last century. Much had happened in these twenty years. Faraday (1791-1867) had discovered electromagnetic induction in 1831, and there was a general awakening to the atomicity of electricity.

By theoretical reasoning Neumann [1.9, 1.10] arrived at the concept of the electrodynamic potential. With the notation of Ampère's theory, this may be expressed as

$$P_{m,n} = \pm \frac{1}{2} i_m i_n \int_m \int_n \frac{\cos \varepsilon}{r_{m,n}} \, dm \, dn \qquad (1.25)$$

It is the mutual potential of two closed circuits, composed of Ampèrian current elements. The double integration involves each pair of current elements twice, but the energy of the pair is only stored once. Neumann took account of this by the factor ½. This has to be born in mind when summing the potential contributions by finite element analysis. Neumann felt uncertain about the sign of the electrodynamic potential until, with the help of the principle of 'virtual work', he was able to derive Ampère's force law from the potential.

Neumann is best remembered for his mutual inductance formula

$$M_{m,n} = \pm \oint_m \oint_n \frac{\cos \varepsilon}{r_{m,n}} \, dm \, dn \qquad (1.26)$$

which arose directly from the electrodynamic potential. The sign of the mutual inductance is also determined by virtual work considerations. In Eq.1.26 the factor of ½ has been dropped because the integrals are now around the two closed circuits. Many precise inductance calculations continue to be based on Neumann's mutual inductance formula. Maxwell incorporated it into field theory. In the process however, he changed its meaning to magnetic flux linkage per unit current.

Comparing the electrodynamic potential, Eq.1.25, with Ampère's force law, Eq.1.24, it will be seen that the dimension of the potential is force times distance, that is energy. Today this potential is called magnetic energy. Any change in the current intensities or the relative distances of the current elements, which increases the potential, requires work to be done on the two circuits. Conversely, if the mutual potential is reduced, energy stored by or in the circuits will be transformed to mechanical work or Joule heat or both. Free energy divorced from matter does not exist in Neumann's theory. The Newtonian electrodynamics treats P as potential energy, depending only on the positions and orientations of the matter elements which enter Ampère's formula. Though both Ampère and Neumann used the term 'current', neither of them ascribed to it kinetic energy, as Maxwell would do later.

Neumann changed his mind about the sign which should be given to his electrodynamic potential. In his first paper [1.9] he defined it as follows:

> "The potential of two closed currents of unit intensity, relative to each other, is the sum of the products of the elements of one current with the elements of the other, each product of the two elements being multiplied with the cosine of the angle of their inclination and divided by their distance."

Following this definition, he used Eq.1.25 with the positive sign. In his second paper [1.10], presented two years later in 1847, he repeated the definition but inserted "the negative half-sum". From then onward he used Eq.1.25 with the negative sign.

In potential theories there have always existed difficulties in agreeing on a universal sign convention. Kellog [1.11] pointed out that the most popular rule was to assign negative potential energy to elements of like sign which attracted each other, and positive potential energy to elements of like sign which repelled each other. Gravitating particles were an example of the former class, and electric charges were an example of the latter class. If matter elements have signs attached to them, they normally represent scalar quantities. Current elements are not of this nature. They have definite directions and therefore they are vectors. Depending on their directions, two current elements sometimes attract and sometimes repel each other. Hence it would seem the potential energy of current elements, and circuits made up of these elements, may sometimes be positive and at other times negative, leaving us to ponder what could be meant by negative energy? We cannot conceive of less than no energy. Consequently, positive and negative potential energy must be two kinds of energy, like positive and negative charge are two kinds of electricity. One kind of potential energy would be associated with attraction and the other with repulsion. Unlike charge, however, the two kinds of energy cannot be neutralized by putting them together.

To see this more clearly, let us now examine two very long straight and parallel wires m and n, as sketched in figure 1.10. In case (a) of that figure they carry currents in the same direction. From experience we know that they will attract each other. By the rules of potential energy they are therefore associated with negative potential energy. Assume an externally applied force F_x tends to increase the separation x and brings about the displacement ∂x by moving n to n'. This external force has to do work and expend an amount of energy equal to $F_x \partial x$. At first it may be thought that this energy is being added to the stored potential energy. This cannot be so, however, because the magnitude of the electrodynamic potential given by Eq.1.25 decreases as a result of the lengthening of the distances between current elements. Not only does the mechanical source sustaining the external force supply energy to the system of conductors, but the potential energy store also gives up energy. What absorbs these two streams of energy? As the currents are assumed to remain constant, no additional Joule heat will be dissipated. Later we will show that, according to Neumann's theory, all of this energy flows to the two electrical sources which maintain the currents. Some of the Joule heat normally furnished by these sources will, during the displacement of one conductor from n to n', be supplied by the potential energy store and the mechanical energy source.

In the case of figure 1.10(b), where the currents flow in opposite directions and the conductors repel each other, the displacement x from n to n' again requires the supply of energy by the mechanical source exerting the external force, but now the magnitude of the stored energy increases because of the shortening of element distances. This opens the possibility of all the energy provided by the mechanical source being stored as potential energy, and the electrical sources maintaining the currents are either not involved in the transaction or they exchange energy with each other.

To appreciate that in a system of conductors positive potential energy does not cancel negative potential energy, we consider three parallel and equidistant wires A, B, and C, in accordance with the model of figure 1.10. If A and B carry current in the same direction, the associated stored energy will be negative, say -P. Let the current in the wire C flow in the opposite direction to the current in B. Then the energy stored between B and C will be positive, that is +P. The forces inside the two sets of conductors do not eliminate each other, and therefore +P-P≠0! The fact that the interaction between A and C adds further positive energy to the system does not change the argument. We have to conclude that positive and negative potential energy are two different kinds of energy which might as well have been named "red" and "green" energy.

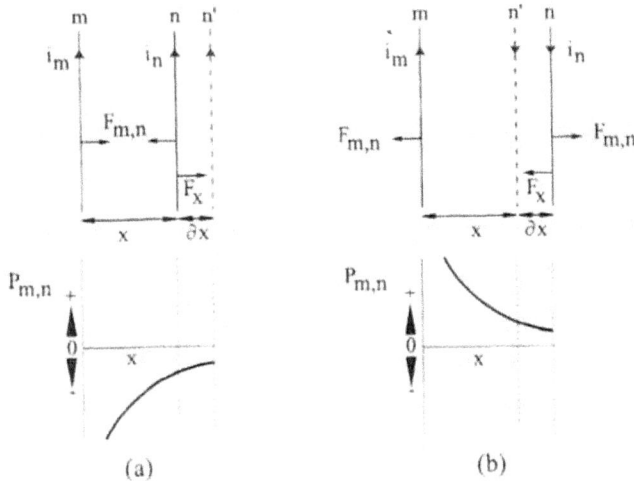

Figure 1.10 : Electrodynamic potential energy of straight and parallel currents

Neumann related the reciprocal force of repulsion or attraction between two circuits m and n to the mutual potential of the circuits and the principle of virtual work by

$$(F_{m,n})_x = - \frac{\partial P_{m,n}}{\partial x} \qquad (1.27)$$

where x denotes a particular direction in which the virtual displacement ∂x takes place. At the same time he chose the negative sign for the potential given by Eq.1.25. The cosine of ε then decides whether, in any particular circuit arrangement, the mutual potential energy turns out to be positive or negative.

When applying Neumann's sign convention to the two conductor arrangements of figure 1.10, it is found that

for figure 1.10(a) $\cos \varepsilon = +1$ (1.28)

for figure 1.10(b) $\cos \varepsilon = -1$ (1.29)

Therefore

for figure 1.10(a) $P_{m,n} = - i_m i_n \oint_m \oint_n \dfrac{1}{r_{m,n}} \, dm \, dn$ (1.30)

for figure 1.10(b) $P_{m,n} = + i_m i_n \oint_m \oint_n \dfrac{1}{r_{m,n}} \, dm \, dn$ (1.31)

This agrees with the rule that attraction is associated with negative energy and repulsion with positive energy. Taking the gradient of the potential energy with respect to x we find

for figure 1.10(a) $\dfrac{\partial P_{m,n}}{\partial x} = + i_m i_n \oint_m \oint_n \dfrac{1}{r_{m,n}^2} \dfrac{\partial r}{\partial x} \, dm \, dn$ (1.32)

for figure 1.10(b) $\dfrac{\partial P_{m,n}}{\partial x} = - i_m i_n \oint_m \oint_n \dfrac{1}{r_{m,n}^2} \dfrac{\partial r}{\partial x} \, dm \, dn$ (1.33)

So we arrive at the interaction force in the specific direction x given by

for figure 1.10 $(F_{m,n})_x = - \dfrac{\partial P_{m,n}}{\partial x} = \pm i_m i_n \oint_m \oint_n \dfrac{1}{r_{m,n}^2} \dfrac{\partial r}{\partial x} \, dm \, dn$ (1.34)

In the case of figure 1.10(a) the force defined by Eq.1.34 is negative, signifying attraction, in agreement with experience. Similarly, for figure 1.10(b), the force becomes positive or repulsion. Hence Neumann's sign convention gives the correct direction of the forces. He extended this proof to the general case of two circuits of any shape. The potential function ultimately adopted by Neumann therefore was

$$P_{m,n} = - i_m i_n \oint_m \oint_n \dfrac{\cos \varepsilon}{r_{m,n}} \, dm \, dn$$ (1.35)

Eq.1.35 will henceforth be used in preference to Eq.1.25.

Neumann did not set out to derive the electrodynamic potential. He discovered it while developing a theory of electromagnetic induction which he based on Ampère's force law. This meant that the potential equation, Eq.1.35, had to be compatible with the force equation, Eq.1.24. The connection lead Neumann to the discovery of the principle of virtual work, as expressed by Eq.1.27. The formal mathematical proof of these facts is very long. It has been fully documented in reference [1.12] and will not be repeated here.

A long forgotten aspect of Neumann's theory is the derivation of electrodynamic turning moments, or mechanical torques, from the electrodynamic potential. Consider two rigid closed circuits carrying currents i_m and i_n, respectively. If circuit N is fixed to the laboratory frame, circuit M will feel torques $(T_{m,n})_x$, $(T_{m,n})_y$, and $(T_{m,n})_z$ on it about arbitrarily chosen cartesian coordinates x, y, and z. Alternatively, if M is fixed, N will feel torques of the same magnitudes, but in opposite directions. Figure 1.11 shows the various torque parameters. The angular displacement about the z-axis is denoted by Ψ_z.

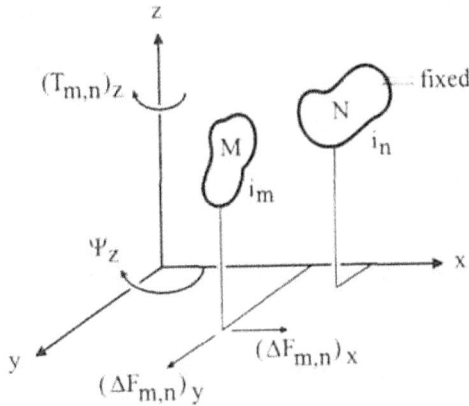

Figure 1.11 Parameters used for electrodynamic torque calculations

The mutual inductance, and therefore the mutually stored potential energy of Eq.1.35 will change when one circuit is turned about an arbitrary axis with respect to the other circuit. This change in stored energy can only be brought about with the aid of a mechanical interaction via mutual torques. With the principle of virtual work, embodied in Eq.1.27 for linear displacements, the corresponding equations for angular displacements must be

$$(T_{m,n})_x = -\frac{\partial P_{m,n}}{\partial \Psi_x} \tag{1.36}$$

$$(T_{m,n})_y = -\frac{\partial P_{m,n}}{\partial \Psi_y} \tag{1.37}$$

$$(T_{m,n})_z = -\frac{\partial P_{m,n}}{\partial \Psi_z} \tag{1.38}$$

Neumann [1.13] did not take the virtual work principle for granted and actually proved Eqs.1.36 - 1.38 from first principles, starting from Ampère's law. This proof has been reproduced in reference [1.12]. In conclusion, Neumann's torque theorem is:

> "The mutual torque between two rigid current-carrying circuits, with respect to any arbitrary axis, is the negative angular gradient of the mutual electrodynamic potential."

In modern field theory Neumann's electrodynamic potential is called stored magnetic energy. Maxwell assumed that this was kinetic energy and therefore it had to be always positive. This has caused difficulties which were absent in Neumann's potential energy treatment.

The potential of Eq.1.35 refers to mutually stored energy between two circuits. The total stored energy should contain additional contributions from the interactions of current element pairs residing in the same circuit. This self-inductance potential and the resulting internal reaction forces were not considered by Neumann. They have become a major point of disagreement between modern relativistic electromagnetism and Newtonian electrodynamics. Neumann was probably held back by the same integration singularities in isolated circuits which have caused considerable controversy in recent years. It will later be shown how these difficulties can be overcome with computer aided finite current element analysis.

In Neumann's theory the stored energy of metallic circuits is associated with the forces of attraction and repulsion between current elements. These elements consist of the substance of the conductor metal, and not the contained electric fluid (conduction electrons). Hence the energy never becomes detached from matter. This is a characteristic of Newtonian mechanics. In contrast to this, modern field theory is based on energy stored in the field and in vacuum. For the magnetic field energy to change, some has to be emitted or absorbed by metal conductors. It requires flying energy transport at the velocity of light and gives rise to many philosophical difficulties which are absent in the Newtonian electrodynamics. For example, there seems to be no satisfactory mechanism that can explain how magnetic field energy is recalled from the far reaches of space when a current is switched off.

Neumann's virtual work method of calculating ponderomotive forces and torques from the change of magnetic energy has survived to this day. It is often preferred to calculations using the Lorentz force. The relevant equations involve Neumann's mutual inductance formula. In many practical arrangements the mutual inductance can be measured with small AC currents. This avoids the most difficult part of the calculations.

It has become common practice to calculate the reaction forces between two parts of the same circuit with Neumann's virtual work concept, but in this case, the mutual inductance is replaced by the self-inductance of the isolated circuit. This procedure gives correct answers, but it cannot be traced back to Neumann.

He shied away from defining the mutual inductance and electrodynamic potential of an isolated pair of current elements. Most of his critical formulae refer to the mutual inductance and stored energy of a pair of complete circuits. The lack of a formula for the mutual inductance between two conductor elements leaves Neumann's theory strangely incomplete. The later chapters of this book fill the gap. The consequences of the mutual inductance formula for two current elements have not been contradicted by experiment.

Neumann's Laws of Electromagnetic Induction

As mentioned previously, Neumann discovered the electrodynamic potential while working on his theory of electromagnetic induction. He started by setting up an elemental law of induction for relative motion between two current elements. For this purpose he assumed that the electromotive force, abbreviated e.m.f., induced in one of the elements, was a function of the current intensity in the other and the Ampèrian force between the elements, provided the element experiencing the induction carried unit current. The latter assumption was made more precise by speaking of the Ampèrian force per unit current in the second element, that is $\Delta F_{m,n}/i_n$, where i_n is the current in the element which experiences the induction. This condition makes the induced e.m.f. in element n, that is Δe_n, independent of the current in this element. Hence the induced e.m.f. in n is finite even when $i_n = 0$. Neumann's induction mechanism is seen to be a one-way process in which the element which carries current is the cause of the induction and the second element experiences the effect. Whether or not there is a back-e.m.f. induced in element m depends on the current i_n. This latter process, however is completely separate from the induction of forward-e.m.f.'s.

There is a clear distinction between Neumann's one-way induction forces and the mechanical forces described by Ampère's law, which are reciprocal forces always involving a two-way process. This difference is reflected by the fact that electromotive forces are measured in volts, while ponderomotive forces are measured in dynes or newtons.

With these postulates, Neumann's elemental law of induction due to relative motion between a current element $i_m dm$ and a conductor element dn can be expressed as

$$\Delta e_n = -v_x \frac{\Delta F_{m,n}}{i_n} \cos \theta_{r,x} \qquad (1.39)$$

where Δe_n is the induced e.m.f. in the conductor element dn shown in figure 1.12. The element dn is taken to be moving with velocity v_x along the arbitrary x-direction relative to the orientation of the inducing element $i_m dm$. $\Delta F_{m,n}$ is Ampère's mechanical force given by Eq.1.24. The angle $\theta_{r,x}$ lies between the distance vector $r_{m,n}$ and the positive x-direction. The negative sign in Eq.1.39 arises from Lenz's law which Neumann [1.9] quotes as follows:

> "If a metallic conductor moves relative to, and in the vicinity of, a galvanic current or magnet, the current induced in the conductor will flow in such a direction that, were the conductor at rest, it would be set in motion in the opposite direction, it being understood that the line of relative motion is fixed."

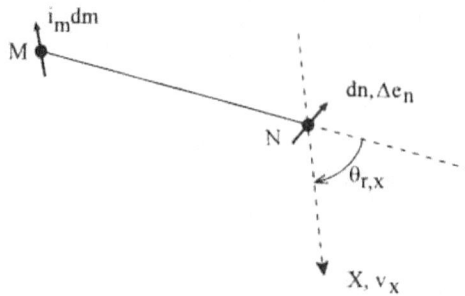

Figure 1.12 : Diagram for Eq.1.39

Neumann treated the proportionality of the induced e.m.f. to the relative velocity v_x and to the inducing current i_m (inside $\Delta F_{m,n}$) as experimentally established facts following Faraday's work and that of others. The elemental law of induction Eq.1.39 is in accord with the Newton-Ampère philosophy of simultaneous mutual matter far-actions. Therefore Δe_n should at all times be proportional to v_x, irrespective of relative acceleration. Neumann felt a little uncertain on this issue. On a number of occasions he referred to the 'stationary state' in which changes in current intensity and relative position of the elements progressed at a rate which was slow compared to the 'velocity of electricity'. As an example of non-stationary phenomena he quoted discharges of capacitors. In order to allow for delays between the cause of induction and its effect, Neumann made a certain dimensional constant a function of time. But then he did not proceed with the analysis of non-stationary phenomena. As can be seen from Eq.1.39, the time-dependent constant was dropped in the later years of Neumann's life. Since then it has been assumed that his law of induction is a law of simultaneous matter interaction.

That the simultaneity of inductive interactions is actually a requirement of the Ampère-Neumann electrodynamics can be seen from the following example. Consider a simple wire circuit connected to the terminals of a battery via a switch. Experience has shown that, when the switch is closed, the current increases smoothly from zero at a rate dictated by the total self-inductance of the circuit. Regardless of how large the circuit may be, the inductive interactions of all wire element pairs appear to spring into action instantaneously when the switch is closed. If the inductive interactions were delayed, the current would, in the first instance, jump to an infinite value, and then decrease to the level dictated by the increasing inductance. However, experiments have never shown a discontinuous jump in the current when the switch is closed, and furthermore, the initial smooth current rise is precisely as predicted from the total self-inductance. Therefore Neumann's original fear that inductive interactions may be delayed was unfounded.

Neumann noticed that the induced e.m.f. was related to the rate of change of the electrodynamic potential. This can be seen by using figure 1.12, and observing that

$$v_x \cos \theta_{r,x} = \frac{dr}{dt} \qquad (1.40)$$

If we assume that the mutual potential of two closed currents as described by Eq.1.35 is the sum of the elemental potential contributions $\Delta P_{m,n}$ from all current element pairs, with one element in either circuit, then Eq.1.39 may be given the form

$$i_n \Delta e_n = \frac{d}{dt} \Delta P_{m,n} \qquad (1.41)$$

The left side of this equation represents power or energy flow to element dn, and the right side gives the rate of change of mutually stored potential energy of the two elements.

Eq.1.39 will be called Neumann's first law of induction. It is an empirical law because it was derived from the experimental facts discovered by Faraday. Even though, as Eq.1.41 indicates, this law could have been derived from the principle of virtual work, the empirical basis is necessary to place it squarely on the foundation of nineteenth century Newtonian electrodynamics.

Interestingly, one experiment has come to light in recent years which can only be explained with Neumann's first law of induction, and not with Maxwell's equations. It is concerned with the operation of railguns and will be discussed in detail in Chapter 5. Maxwell claimed that he had incorporated Neumann's theory into his equations, and particularly in what has become known as Faraday's law of induction. It is now evident that this marriage of the far-action theory with field-contact action was less than perfect.

To establish his second law of electromagnetic induction, Neumann went to great length to show that an equation like Eq.1.41 also applies to two complete circuits m and n. This can be written

$$i_n e_n = \frac{dP_{m,n}}{dt} \qquad (1.42)$$

Substituting for the electrodynamic potential from Eq.1.35, the total e.m.f. induced in the circuit n becomes

$$e_n = - \frac{d}{dt} \oint_m \oint_n \frac{i_m \cos \varepsilon}{r_{m,n}} dm \ dn \qquad (1.43)$$

or

$$e_n = - \frac{d}{dt} i_m M_{m,n} \qquad (1.44)$$

where $M_{m,n}$ is the mutual inductance of the two circuits given by Eq.1.26. The step from Eq.1.43 to Eq.1.44 may be taken as the definition of Neumann's mutual inductance in terms of matter interactions rather than magnetic flux linkage. Modern textbooks on electromagnetism call Eq.1.44 Faraday's law because the product of current and mutual inductance is the mutual flux linkage between the two circuits. Without wishing to take away anything from the important experimental achievements of Faraday, we will describe Eq.1.44 as Neumann's second law of electromagnetic induction. This corrects the historical record.

When computing mutual inductances with Neumann's formula, Eq.1.26, it is necessary to assign directions of current flow to the two closed circuits, as only this will make the angle ε unique for every element pair. The $\pm$ sign of Eq.1.26 acknowledges this uncertainty with respect to the current directions. Reversing the direction of current in one of the circuits will not change the magnitude of the mutual inductance but reverses its sign.

Provided the conductor elements belong to two closed circuits, it follows from Eq.1.42 that the elemental e.m.f. may be expressed as

$$\frac{\Delta e_n}{dn} = - \frac{d}{dt}\left(\frac{i_m \ dm \ \cos \varepsilon}{r_{m,n}} \right) \tag{1.45}$$

The quantity inside the bracket turns out to be the magnetic vector potential of the current element $i_m dm$ at point N, the center of the conductor element dn. Neumann wrote his papers before vector analysis was invented and he did not mention the magnetic vector potential. Denoting the vector potential by $\vec{A}$, Neumann's second law of induction may be stated as

$$\frac{\Delta \vec{e}_n}{dn} = - \frac{d}{dt} \Delta \vec{A}_{m,n} \tag{1.46}$$

where the vector potential is given by

$$\Delta \vec{A}_{m,n} = \frac{i_m \ d\vec{m}}{r_{m,n}} \quad ; \quad \Delta \vec{A}_{n,m} = \frac{i_n \ d\vec{n}}{r_{m,n}} \tag{1.47}$$

The vector potential is not a reciprocal interaction parameter because it involves only one current element at a time. Therefore

$$\Delta \vec{A}_{m,n} \neq \Delta \vec{A}_{n,m} \tag{1.48}$$

As a consequence of Eq.1.35, the mutual electrodynamic potential of a pair of current elements, belonging to separate closed circuits, is

$$\Delta P_{m,n} = - i_m \ i_n \ \frac{\cos \varepsilon}{r_{m,n}} \ dm \ dn \tag{1.49}$$

and using Eq.1.47 this is equal to the scalar products

$$\Delta P_{m,n} = i_m \vec{dm} \cdot \Delta \vec{A}_{n,m} = i_n \vec{dn} \cdot \Delta \vec{A}_{m,n} \tag{1.50}$$

Eq.1.50 reveals just how closely the magnetic vector potential is related to Neumann's electrodynamic potential.

Closed conduction currents may induce e.m.f's in open-circuited conductor sections. An example is the combination of a loop antenna with a dipole antenna. This problem was examined by Neumann. Consider a conductor n consisting of just a single element dn. In order to obtain the action of the closed current i_m on dn, using Eq.1.41 and Eq.1.49 it follows that

$$\Delta e_n = - \ dn \ \frac{d}{dt} \left(i_m \oint_m \frac{\cos \varepsilon}{r_{m,n}} \ dm \right) \tag{1.51}$$

The angle function in Ampère's force law, Eq.1.24, is written as

$$f(\alpha, \beta, \varepsilon) = (2 \cos \varepsilon - 3 \cos \alpha \cos \beta) \tag{1.52}$$

But Neumann had proved that, when one of the circuits is closed

$$dn \oint_m \frac{2 \cos \varepsilon - 3 \cos \alpha \cos \beta}{r_{m,n}} \ dm = dn \oint_m \frac{\cos \varepsilon}{r_{m,n}} \ dm \tag{1.53}$$

and thus in this case

$$f(\alpha, \beta, \varepsilon) = \cos \varepsilon \tag{1.54}$$

It is this restricted angle function which is being used in Eq.1.51.

If the conductor n consists of more than one element and extends from n_1 to n_2 the e.m.f. induced in this length of conductor is

$$e_n = \frac{d}{dt} \left[i_m \int_{n_1}^{n_2} \oint_m \frac{\cos \varepsilon}{r_{m,n}} \ dm \ dn \right] \tag{1.55}$$

Neumann's theory of electromagnetic induction, pertaining to metallic conductors, has survived in modern field theory, however the words around the formulae have changed. Where Neumann spoke of interacting conductor elements and complete circuits modern physics now talks of magnetic flux linkage. The e.m.f. per unit length has become the electric field intensity, and so on. The flux linkage idea breaks down when one of the circuits is unclosed. Neumann's method, on the other hand, can deal with the e.m.f. in an unclosed conductor, as has been shown with Eq.1.55.

Since the electrodynamic potential was derived from Ampère's force law, and since this potential largely survives in field theory, one might expect Maxwell's equations to contain Ampère's force law, but in fact they do not. Maxwell [1.8] himself was aware that field theory does not contain a force law. He strongly endorsed Ampère's law but thought the Grassmann formula, to be discussed in the next section, would do equally well. The Grassmann law has become the magnetic component of the Lorentz force acting between two moving charges. In modern electromagnetism this has taken over the function of Ampère's law.

Classical Newtonian physics was based on the pillars of three empirical force laws, those of Newton, Coulomb, and Ampère. They were all simultaneous far-action laws and ushered in the first two centuries of quantitative science. Modern physics has made a complete break with far-actions. The first step in this direction was taken by Maxwell (1831-1879), however before then, the far-action electrodynamics had been developed in other directions, as discussed in the following section.

Grassmann's Force Law

Both Ampère and Neumann had concerned themselves not only with the interaction of linear currents in metallic conductors, but also with the mechanical and electromotive forces between magnets and in mixed systems containing magnets and current circuits. Ampère's concept of the magnetic molecule has had lasting value. Neumann showed the equivalence of a current carrying circuit and a magnetic shell bounded by the circuit. Our book does not deal with the behaviour of magnetic materials and concerns itself solely with the interaction of electric currents in non-magnetic metals. In this restricted sense, electrodynamics is the science of metallic current elements.

The force law for two current elements which has been used almost to the exclusion of all others during the past eighty years was first proposed by Grassmann (1809-1877) in 1845 [1.14], the same year Neumann published his theory of induction. Grassmann's is an unsymmetrical law and therefore has to be stated by two equations. One is for the force on element dm, to be written $\Delta \vec{F}_m$, and the other for the force ΔF_n acting on element dn. In vector form, but otherwise with the previously employed notation, these two equations are

$$\Delta \vec{F}_m = \frac{i_m i_n}{r_{m,n}^2} \, \vec{dm} \times (\vec{dn} \times \vec{a}_{r,m})$$

$$\Delta \vec{F}_n = \frac{i_m i_n}{r_{m,n}^2} \, \vec{dn} \times (\vec{dm} \times \vec{a}_{r,n})$$

(1.56)

where the direction of the unit distance vectors $\vec{a}_{r,m}$ and $\vec{a}_{r,n}$ is along the line connecting the elements and pointing toward the element at which the force is being determined. Grassmann considered Eq.1.56 to be a far-action law between two elements of matter, strictly in the Newtonian sense.

To see the relationship to the Ampère-Neumann electrodynamics it is best to resolve the triple vector products of Eq.1.56 according to

$$\vec{A} \times (\vec{B} \times \vec{C}) = \vec{B}\,(\vec{A} \cdot \vec{C}) - \vec{C}\,(\vec{A} \cdot \vec{B}) \qquad (1.57)$$

Applying Eq.1.57 to Eq.1.56 results in

$$\Delta \vec{F}_m = \vec{dn}\,\frac{i_m\,i_n\,dm}{r_{m,n}^2}\cos\alpha_m - \vec{a}_{r,m}\,\frac{i_m\,i_n\,dm\,dn}{r_{m,n}^2}\cos\varepsilon$$

$$\Delta \vec{F}_n = \vec{dm}\,\frac{i_m\,i_n\,dn}{r_{m,n}^2}\cos\alpha_n - \vec{a}_{r,n}\,\frac{i_m\,i_n\,dm\,dn}{r_{m,n}^2}\cos\varepsilon \qquad (1.58)$$

The angles α_m and α_n must not be confused with α and β of Ampère's force law, Eq.1.24, but ε is the same in both laws. Figure 1.13 should help to illustrate the respective angle conventions.

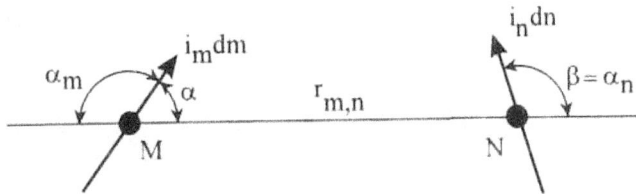

Figure 1.13 : Angle conventions for Ampère's and Grassmann's force laws, Eqs.1.24 and 1.58

According to Eq.1.56 a pair of current elements do neither attract nor repel each other. Each experiences a force perpendicular to itself which has its cause in the existence of the other. The transverse force lies in the plane containing the element in question and the line connecting both elements.

Grassmann also pointed out that, as a result of Ampère's rule, the interaction of two current elements always reduces to a two-dimensional problem. With respect to figure 1.13 this means that if we wish to determine the force on element dm we need only consider that component of the other element which lies in the plane of dm and $r_{m,n}$. In compliance with Ampère's rule, the component of the other element which is perpendicular to this plane produces no force on the element dm. Therefore the interacting components of the two current elements and the Grassmann forces all lie in the same plane.

The abandonment of mutual attraction and repulsion between matter elements of electric conductors, and the violation of Newton's third law which this entailed, signalled the end of Newtonian physics. The Grassmann and Lorentz force laws required a new mechanics which was to become that of the theory of special relativity. Regardless of velocity of light issues, special relativity was already inherent in the electrodynamics of Grassmann and Lorentz.

In the expanded form of Grassmann's law, Eq.1.58, the second term is a Newtonian term of repulsion or attraction. The remaining term is a force acting in the direction of the other current element. It is this term which violates Newton's third law. It will be called the relativistic term of the Grassmann or Lorentz force. Whittaker [1.6] and others have shown that when the Grassmann force on, say, dm is summed over a closed circuit n, all the relativistic contributions add up to zero. Only the force contributions made by the Newtonian term survive, and they agree with Ampère's law. Hence if in the Grassmann electrodynamics forces are calculated which involve one or more closed metallic circuits, we automatically, and in most cases unknowingly, slip back into the mathematics of the Newtonian electrodynamics. This mathematical deception has confused many field theoreticians.

Grassmann gave the magnitude of the perpendicular force acting on dn as

$$\Delta F_n = i_m \, i_n \, \frac{dm \ dn}{r_{m,n}^2} \sin \theta \qquad\qquad (1.59)$$

where θ is the angle of the Biot-Savart law, Eq.1.2, but with the ds-element replaced by the dm-element. Figure 1.14 depicts the connection between the Grassmann and Biot-Savart laws. The Biot-Savart law gives the magnetic field strength $\vec{dH}$ at N due to the dm current element. Therefore

$$\Delta \vec{F}_n = i_n \, \vec{dn} \times \vec{dH} \qquad\qquad (1.60)$$

This last equation clearly reveals that the Grassmann force is actually the magnetic component of the Lorentz force of modern field theory.

It is rather surprising to find that Grassmann had enunciated his law twenty years before Maxwell wrote his field equations, but at the time only Faraday was speaking of magnetic flux. Grassmann was a mathematics teacher at a German high school. Grassmann's great achievement as a mathematician was the introduction of vector calculus. There is some suspicion that he proposed his new electrodynamics [1.14] mainly in order to have a good application for vectors. He certainly achieved this with Eq.1.56. The Lorentz force expression made its appearance in the 1890s, fifty years after Grassmann's paper. Lorentz was led to Eq.1.60 by his theory of electrons [1.15].

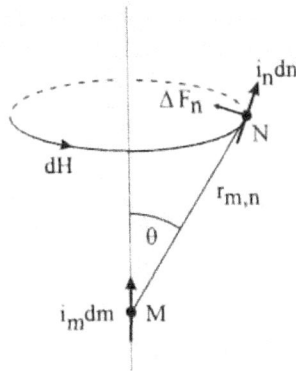

Figure 1.14 : Derivation of Grassmann's force law from the Biot-Savart law

On Grassmann's own authority, his investigation was prompted by two objections to Ampère's force law. He considered attraction and repulsion to be an arbitrary assumption, and also could see no reason why current elements should behave like gravitating and charged particles which were scalar quantities while current elements were vectors.

As far as his second objection to Ampère's law was concerned he said [1.4]:

"The complicated form of this formula arouses suspicion, and the suspicion is heightened when an attempt is made to apply it. If, for example, the simplest case is considered, in which the circuit elements are parallel, so that $\varepsilon = 0$ and $\alpha = \beta$, the Ampère expression becomes

$$\frac{i_m \, i_n \, dm \, dn}{r_{m,n}^2} \, (2 - 3 \cos^2 \alpha)$$

from which it appears that, when $\cos^2\alpha$ is equal to 2/3 or, which comes to the same thing, $\cos 2\alpha$ is equal to 1/3, that is the position of the midpoint of the attracted element lies on the surface of a cone whose apex is at the attracting element, and whose apex angle is arccos(1/3), there is no interaction; while for smaller angles there is repulsion, and for larger ones attraction. This is such an unlikely result, that the principle from which it is derived must come under the gravest suspicion and with it the supposition that the force in question must show an analogy with all other forces."

Ampère's force reversal which takes place when a current element, held parallel to itself, describes a circle around another element is plotted on figure 1.15. The mutual Ampère force varies from one arbitrary unit of repulsion at $\alpha=0$ or $180°$ to two units of attraction at $\alpha=90°$ or $270°$. In each quadrant there exists an angular position for which the force is zero, and changes from attraction to repulsion, or vice versa. Grassmann did not like this unexpected variation of the elemental force with angular position, but was unable to provide an argument which proved Ampère's force law to be wrong.

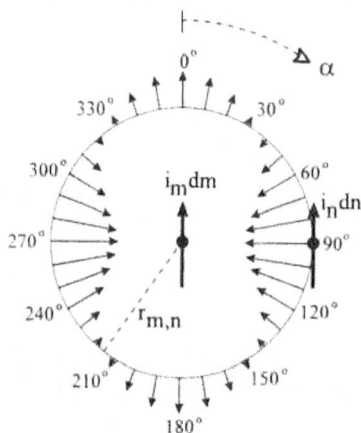

Figure 1.15 : Polar diagram of the Ampère force between two parallel current elements with a constant distance of separation

Lorentz established that Grassmann's formulae could be treated as empirical laws for electric charges drifting in vacuum. To do this he changed the definition of a current element. In the Ampère-Neumann electrodynamics this was a piece of matter of length dl, carrying an electric current i, such that the current element could be expressed as i.dl. Lorentz made the current element the product of an electric charge q and a relative velocity v, which could be written qv. The change in the definition of the current element had major physical consequences which will be discussed in later chapters.

Grassmann had no experimental results in 1845 with which he could support his force law. Instead he used mathematical and verbal arguments. His mathematical treatment is difficult to understand as he was still groping for an easy method of handling vectors.

The verbal logic which Grassmann put forward is interesting and worth summarizing. He had recognized that an infinitely long current behaves like a closed loop in its interaction with other currents. That is to say, the force which an infinitely long current will exert on an element of another current is always perpendicular to the latter, as proved by the wire-arc experiment of Ampère. Grassmann then extended this principle to an angle-current (Winkelstrom). This is an infinitely long current forming the two arms of an angle, the current coming from infinity in one straight conductor and returning to infinity in another. Subsequently he relied on the force on any current element lying in the plane of the angle-current to be perpendicular to the element, whatever its position or orientation in that plane. As Ampère had done, Grassmann assumed the wire to consist of a very large number of straight and short matter elements. Each element could be thought of as lying to one side of the apex of the angle-current. The idea is further explained by figure 1.16 which deals with a clever arrangement in which the closed circuit ABCDE consists merely of five elements. Element AB is part of the angle-current aAb and the element BC of angle-current bBc, and so on. It will be appreciated that each of the infinite rays a, b, c, d, and e carry outgoing and incoming currents of the same magnitude superimposed on each other and therefore, in fact, carry no current at all. With this mental picture Grassmann considered the force exerted on an external current element, $i_m dm$ in figure 1.16, and particularly the interaction of this external current element with any of the elements of a closed circuit such as ABCDE. He was convinced that each of the circuit elements would, independently of the others, generate a perpendicular force on the external element, because each of the elements of ABCDE was also part of a separate angle current. The algebraic sum of the perpendicular forces on $i_m dm$ was then the total force exerted by the circuit ABCDE on the external elements. This was the basis on which Grassmann justified the directions in his force law Eq.1.56.

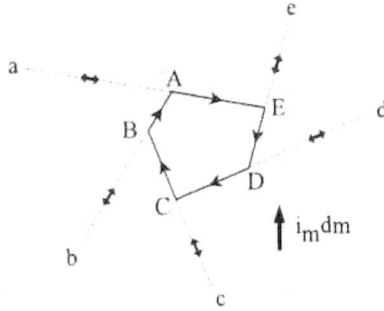

Figure 1.16 : Grassmann's angle-current representation of a closed circuit

Grassmann's argument requires substantiation in two respects. It is not clear whether forward and reverse currents can be superimposed to cancel their effects. Secondly, the five current elements of figure 1.16 cannot be fitted together as shown, for every element must have a volume, and the volumes would overlap.

Grassmann accepted Ampère's proof of the inverse square of distance relationship and the proportionality of the force to the product of the current intensities and element lengths. The $\sin \theta$ of Eq.1.59 stems from the Biot-Savart law, but Grassmann arrived at it independently. The agreement between Grassmann's law, Eq.1.59, and the expansion of the triple vector product Eq.1.58 may be shown with figure 1.17. In this diagram $i_n dn$ is the resolved component of the general element at N in the plane of $i_m dm$ and $r_{m,n}$. Furthermore if

$$k = i_m \, i_n \, \frac{dm \; dn}{r_{m,n}^2}$$

then $k \cos \varepsilon$ is the magnitude of the Newtonian vector of Eq.1.58 and $k \cos \alpha_n$ is the magnitude of the relativistic vector of the same equation. Applying the sine rule to the force triangle of figure 1.17 gives

$$\frac{\Delta F_n}{\sin \theta} = \frac{k \cos \alpha_n}{\sin ((\pi/2) - \alpha_n)}$$

Therefore

$$\Delta F_n = k \sin \theta$$

which proves the magnitude equality of Eqs.1.58 and 1.59.

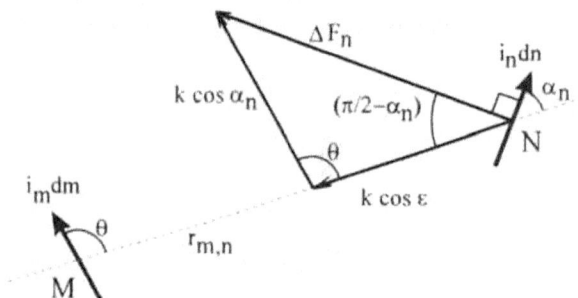

Figure 1.17 . Force triangle used to prove equality of Eqs. 1.58 and 1.59

Grassmann's new electrodynamics had little impact on his contemporaries. It would now be completely forgotten but for the fact that it fitted in well with the 20th century field, relativity, and electron theories. Although it was based on many of Ampère's ideas and experiments, his rejection of the concept of balanced action and reaction between each pair of current elements meant that Grassmann's law was not a part of the Newtonian electrodynamics.

Weber's Force Law and Electrodynamic Potential

Finding mathematical laws capable of quantifying Faraday's 1831 discovery of electromagnetic induction took much longer than Ampère's deduction of a force law from the discovery of electromagnetism. No less than fourteen years elapsed before Neumann in Köenigsberg published his laws of induction. Other scientists had been studying this problem at the same time. Among them were Fechner [1.16] and Weber [1.17] in Leipzig. Neumann had derived his theory of electromagnetic induction without hypothesis as to the nature of the electric fluid. He did, however, have to invent a new force, which he called the electromotive force, to distinguish it from the ponderomotive force which moved the metallic conductor.

The Leipzig school apparently appealed to only one kind of force on charged particles that possessed mass and constituted the electric fluid. Electrolysis had revealed the existence of discrete charged particles in an electric current, yet Fechner and Weber did not know of the rigid connection of positive charges (ions) to the metal lattice. That not all electrodynamic forces in a metal are mechanical forces on the lattice must have something to do with the two types of bonds that (a) exist between positive and negative charges and (b) between charges and the solid body. This issue of bonding between charges and ponderable matter has still not been satisfactorily resolved even at the end of the twentieth century.

In its experimental consequences, Weber's work, in the end, added little to that of Coulomb, Ampère, and Neumann. Weber based his force law on the same mutual simultaneous far-action principle underlying all of the Newtonian electrodynamics. What stands out in Weber's writings is that he was the first to take notice of the atomicity of electricity. He formulated a new model of the metallic current element in terms of mobile charged particles. As the following chapters will show, the search for a satisfactory model of the Ampèrian current element, compatible with all of solid state physics, is still in progress today. Weber's theory will be reviewed to illustrate some problems which have to be faced in order to find the correct microscopic current element model.

At the instigation of Fechner, Weber searched for a force expression which was mathematically equivalent to Ampère's well-proven law. Accepting the two-fluid model of electric current, Weber used positive and negative charges streaming in opposite directions, past each other, through and along current elements. Then he hypothesized that the interaction forces between the charged mass particles would not only depend on their relative positions, but also on their relative motions. From the start he intended to unify electrostatics with electrodynamics, and therefore Weber's force law had to include Coulomb's law, Eq.1.3, for two charges at relative rest with respect to each other.

Weber proposed the following empirical laws for the force $\Delta F_{e,e'}$, between two electric charges e and e', and the mutual potential $P_{e,e'}$, associated with this force

$$\Delta F_{e,e'} = \frac{e\,e'}{r^2}\left[1 - \frac{1}{2c^2}\left(\frac{dr}{dt}\right)^2 + \frac{r}{c^2}\frac{d^2r}{dt^2}\right] \tag{1.61}$$

$$\Delta P_{e,e'} = \frac{e,e'}{r}\left[1 - \frac{1}{2c^2}\left(\frac{dr}{dt}\right)^2\right] \tag{1.62}$$

where r is the distance between the charges, t is time, and c is a dimensional constant. The Ampère-Neumann electrodynamics was formulated in fundamental electromagnetic units (e.m.u.). The Weber equations, in contrast, are given in fundamental electrostatic units (e.s.u.) because they contain Coulomb's law which defines the e.s.u.-system of units. According to this system, the product of two electrostatic units of charge divided by their distance (in centimeters) squared gives the interaction force in dynes. Therefore the factor outside the main bracket in Eq.1.61 has the dimension of a force. The terms inside the bracket have to be dimensionless numbers. This means c must have the dimension of velocity. Weber went on to show that if Eq.1.61 was to be in agreement with Ampère's force law, Eq.1.24, when the current in each element is carried by a single positive electric charge travelling with constant velocity with respect to the metallic conductor element containing it, and furthermore if this law was also to agree with the results of Ampère's experiments, then the constant had to have the value $c = 3 \times 10^{10}$ cm/s. This constant became known as the velocity of light and it always emerges when the laws of electrostatics are combined with those of electrodynamics.

It was Weber's research into the fundamental measures of electromagnetism which revealed the importance of the c-factor. Using the same unit of force in the laws of Coulomb and Ampère, he calculated that 3×10^{10} electrostatic units of charge had to pass through the current element every second to represent the flow of one electromagnetic unit of current. The same rate of charge transfer is obtained when one charge passing through the current element travels at the velocity of light. This is how the velocity of light made its first appearance in the literature and Newtonian electrodynamics. In 1857 Kirchhoff [1.18] proved with Weber's law that electrical disturbances travel with the velocity of light along transmission lines. Readers of modern textbooks are often misled to believe that Maxwell was the first to discover the role which the velocity of light plays in electromagnetism.

Weber attributed no particular importance to c, however today it appears truly astonishing that the velocity of light should have sprung up in a simultaneous far-action theory such as his. Although the charges to which Eq.1.61 relates, move relative to each other and their distance r is a function of time, the forces of repulsion or attraction between the charges are assumed to change simultaneously with r. The formula does not allow for an energy propagation delay which could be linked to the velocity of light.

Weber [1.17] proved in detail how his force law, Eq.1.61, can be transformed to Ampère's force law, Eq.1.24. His transformation is a long mathematical process and teaches little. Of course Weber could not have guessed the form of Eq.1.61. In fact he derived it from the work of Coulomb and Ampère. His method of derivation is very instructive and worth repeating in brief outline. A more complete derivation can be found in reference [1.12].

Weber published his force law in 1846, the year between the two Neumann memoirs [1.9, 1.10], and he was obviously not aware of Neumann's researches on induction, to which he referred extensively in later years. Weber began his force law deduction as follows:

"To lay down a guideline for this study, which is based on experience, we consider three specific facts resting partly on direct observation and partly on the indirect measurements underlying Ampère's fundamental law.

(1) The first fact is that two current elements lying on the same straight line either repel or attract each other, depending on whether their currents flow in the same or opposite directions.

(2)   The second fact is that two parallel current elements perpendicular to the line joining them either attract or repel each other, depending on whether their currents flow in the same or opposite directions.

(3)   The third fact is that a current element, which lies on a straight line with a wire element, induces similarly directed or opposed current, depending on whether its own current intensity decreases or increases.

These three facts are not direct results of experiments, because the action of an element on another cannot be observed, but they accurately correspond to observed phenomena to the extent that they almost have the same validity. The first two facts are already incorporated in Ampère's basic formula of electrodynamics and the third has been added by Faraday's discovery (of induction)."

Figure 1.18 depicts Weber's model of two interacting current elements. Each element need only contain one positive and one negative charge. The two charges in each element move toward each other, along the line of the element, with velocity v relative to the metal. They are allowed to pass each other without appreciable deviation because, as Weber explained, we are not dealing with the actual happenings in the conductor but only with an action at a distance theory in which the charges are treated as if they could pass each other on a line. The current intensity of the element is taken to be ev, where e is the positive charge.

Four Coulomb-type interactions have to be considered, two of which are repulsions of like charges and the remaining two are attractions of unlike charges. All four sets of forces act along r, the line connecting the center points M and N of the two elements. All that was known about the forces at Weber's time was their strengths as given by Coulomb's law, Eq. 1.3, for the case where they are at rest with respect to each other. Weber deemed it probable that relative motion between the charges would modify the actions, and Coulomb's law would give the limiting value of the forces when the relative velocities tended to zero. He considered it to be his task to determine the departure from Coulomb's law as a function of the relative motion between the charges.

Of the three facts on which Weber claimed he had built his theory, (1) and (2) referred to Ampère ponderomotive forces on the conductor metal, but (3) involved a Neumann-type electromotive force on charge. Weber convinced himself, however, that the total force experienced by the metal of the current element was the vector sum of all the electric forces on charges within it by charges located elsewhere. Since charges cannot penetrate the surface of the conductor, forces on charges which are transverse to the wire axis will be directly transmitted to the metal. This mechanism of force transmission to the wire material is not available to electromotive forces along current elements to generate Faraday's induced currents. With regard to these latter forces Weber seemed to argue that, since they are only of a transient nature and their magnitude is related to the ratio of the mass of the moving charges to the much greater mass of the stationary metal, they may be ignored in the calculation of mechanical forces on the metal. Weber apparently did not appreciate that the electromotive forces in a homopolar generator (Faraday disk) are not of a transient nature.

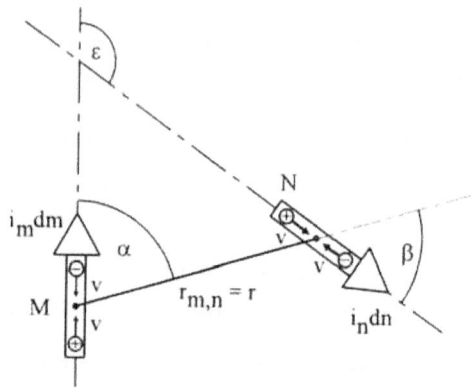

Figure 1.18 . Weber's model of two current elements

The lack of a credible mechanism of force transmission from freely moving charges to the body of the metal is the weakest point of Weber's electrodynamics. This is particularly true for longitudinal electrodynamic forces which generate tension in wire conductors.

The relative velocity of two charges separated by the distance r is (dr/dt). This differential coefficient is positive when the charges move apart and negative when they come together. In order to make the magnitude of that component of the inter-particle force, caused by relative motion, the same for approaching and receding motions, Weber assumed the second term of his force law, Eq.1.61 to be proportional to the square of the relative velocity.

According to Weber's rule of the stronger velocity-dependent interactions of charges moving in the same direction, the second term, furthermore, had to reduce the Coulomb forces. This determined its negative sign. The second term now had to be proportional to $-(dr/dt)^2$. The full mathematical analysis leading to the $(1/2c^2)$ factor is outlined in reference [1.12].

The third term of Weber's force law followed from the knowledge that side-by-side elements exert larger forces on each other than in-line elements. This refers to fact (2) of the three facts on which Weber's theory was based. To account for the stronger interaction of parallel elements, which are arranged perpendicular to the line joining them, it was necessary to call upon the relative acceleration d^2r/dt^2 contained in the third term of Weber's force law, Eq.1.61. The detailed mathematical steps of this long deduction can also be found in reference [1.12].

For two current elements $i_m dm$ and $i_n dn$, the electrostatic interaction will be zero, and Weber's force law reduces to the $(dr/dt)^2$ and the d^2r/dt^2 terms. It is only when considering the interaction force between any two charges e and e' that the full formula Eq.1.61 is required. In this formula the force is given in dynes when e and e' are inserted in e.s.u. of charge, r is in centimeters, and t in seconds. As explained before, the velocity c in Eq.1.61 is the velocity of light in vacuum.

Writing Weber's electrodynamic potential Eq.1.62 of two charged particles as

$$\Delta P_{e,e'} = \frac{e\,e'}{r} - \frac{e\,e'}{2\,r\,c^2}\left(\frac{dr}{dt}\right)^2 \tag{1.63}$$

it is seen to be the difference between the electrostatic potential and a term containing the relative velocity between the charges. The term involving the square of the relative velocity is akin to a quantity of kinetic energy given by

$$\frac{1}{2}\,m_e\,v_r^2$$

where m_e is some non-material electromagnetic mass and v_r the relative velocity between the charges. Weber's potential thus defines the electromagnetic mass as

$$m_e = \frac{e\,e'}{r\,c^2} \tag{1.64}$$

which leads to the following mass-energy relationship

$$m_e\,c^2 = \frac{e\,e'}{r} \tag{1.65}$$

This represents the first time that a term of the form mc^2 appeared in the scientific literature, and should be compared with Einstein's mass-energy equation ($E=mc^2$). The left-hand side of Eq.1.65 will be recognized as being similar to the magnetic field energy (with a different electromagnetic mass) of relativistic electromagnetism which is being transported by the Poynting vector, and yet the right side is a Newtonian potential energy.

Because of the inclusion of some kinetic energy in Weber's potential, the inter-particle force is not simply the negative gradient of the potential, as it was in Neumann's virtual work theory, but the Lagrange-force defined by the differential operator

$$\left[-\frac{\partial}{\partial r} + \frac{d}{dt}\frac{\partial}{\partial v_r} \right]$$

that is

$$\Delta F_{e,e'} = \left[-\frac{\partial}{\partial r} + \frac{d}{dt} \frac{\partial}{\partial v_r} \right] \left\{ \frac{e \, e'}{r} \left(1 - \frac{1}{2c^2} v_r^2 \right) \right\}$$

$$= \frac{e \, e'}{r^2} \left\{ 1 - \frac{1}{2c^2} \left(\frac{dr}{dt} \right)^2 + \frac{r}{c^2} \frac{d^2r}{dt^2} \right\} \qquad (1.66)$$

which is the Weber force law. It should be noted that, just as with photons, Weber's electromagnetic rest mass is zero.

The gathering of experimental facts in confirmation of Ampère's longitudinal forces led to a revival of the Weber electrodynamics in the decade of the 1980s. The hope was that a far-action electrodynamics would emerge which was based on forces between moving charges. Ampère's electrodynamics had offered no clues as to how the longitudinal forces could eventually be justified at the atomic level.

Modern Weber theorists are taking note of the fact that we now know that positive charges in the metal are frozen to the lattice, preventing them from moving relative to the conductor metal. The unusual Fechner hypothesis of the counter-directional flow of positive and negative charges, while the current consists only of the motion of the positive charges, has had to be dropped. Lorentz's electron theory of metals allows for the flow of conduction electrons. This is now defined as a negative current.

Assis [1.19] has shown that the fixed positive charges and mobile negative charges do not in any way change the appearance of Weber's force law, Eq.1.61 and electrodynamic potential, Eq.1.62. The atomic bonds which hold the lattice ions in place form an ideal means of transferring forces on positive charges to the body of the metal. No such bonds are available to transfer the forces on the conduction electrons to the metal lattice. These same electrons must also be free to respond to induced electromotive forces predicted by the Weber electrodynamics. Therefore if it seems impossible for the Weber theory to account physically for longitudinal electrodynamic tension in wires, then this difficulty arises from Weber's current element model.

Weber's theory, as well as Lorentz's modern electron theory [1.15], treat the current element as the product of a charge multiplied by a velocity. Ampère's law contains no velocity. A number of the experiments to be discussed in Chapter 2 suggest that all of Ampère's current element is anchored on the lattice site. It indicates that the Ampèrian current element may be some form of electric or magnetic dipole.

Kirchhoff's Circuit Theory

Gustav Kirchhoff (1824-1887) was Franz Neumann's most illustrious pupil. As a young man, and before Maxwell published any of his field theory papers, Kirchhoff developed what has become known as 'circuit theory'. This has proved most useful for electrical engineers and promises to continue to do so for many more years.

Today it is forgotten that circuit theory has its roots in the Newtonian electrodynamics of Coulomb, Ampère, Neumann, and Weber. It is, therefore, an action at a distance theory which, on close examination, disagrees with various aspects of modern field theory. For example, Kirchhoff [1.20-1.22] proved with circuit theory that voltage and current waves travel along wires with the velocity of light. This remarkable fact arose from multiple inductive and capacitive far-actions of huge numbers of current and conductor elements. Kirchhoff was in fact the first to derive delays in the transmission of electrical disturbances along conductors with a many-body interaction model. The strange aspect of the complex far-action calculations is that they predict the same time delays as the simple energy transport model of field theory.

In field theory it is asserted that the transmission of electrical signals along a two-wire line requires the flight of free electromagnetic energy between the wires. Moreover, this energy must travel with the velocity of light. One may speculate that the energy transport model represents the true physical state of affairs, and far-action theory is simply an abstract mathematical framework which furnishes the same signal propagation velocity. Should it be shown, however, by experiment that the free energy transport between the wires is fiction, then we would have little choice but to endow the far-action mechanism with a degree of physical reality. The experimental resolution of this question is an important aspect of our book.

First and foremost, circuit theory clarified the concepts of voltage and current in conjunction with Ohm's law and the definition of electrical resistance. In addition, the capacitance parameter took over the science of electrostatics, and the inductance parameters (self and mutual) did the same for electrodynamics. These are the reasons why with the three lumped circuit parameters of resistance, capacitance, and inductance, in addition to Kirchhoff's laws of the distribution of voltages and currents in electric networks, we can solve almost any problem in electrical engineering which does not involve the radiation of electromagnetic energy.

This completes the review of the Newtonian electrodynamics as it evolved in the nineteenth century. Before field theory became fully accepted there was a period in which retarded, and even advanced, potentials found favour. Retarded and advanced potentials are really the science of flying forces. The logistics of this very complex force transport however, was never satisfactorily resolved. Hence at the end of the twentieth century we are left with the Newtonian model of action at a distance, and field contact physics, based on the flight of energy, and championed by Maxwell, Lorentz, and Einstein.

Chapter 1 References

1.1   Bern Dibner, *Oersted and the discovery of electromagnetism*, Blaidsdell Publishing, New York, 1962.

1.2   H.C. Oersted, "Experiments on the effect of a current on the magnetic needle", Annals of Philosophy, Vol.16, 1820.

1.3   P. Hammond, "Andre-Marie Ampère: the Newton of electricity", Journal IEE, p.274, 1961.

1.4   R.A.R. Tricker, "Ampère as a contemporary physicist", Contemporary Physics, Vol.3, p.453, 1962.

1.5   R.A.R. Tricker, *Early electrodynamics*, Pergamon Press, London, 1965.

1.6   E. Whittaker, *A history of the theories of aether and electricity*, Thomas Nelson, London, 1951.

1.7   A.M. Ampère, "La determination de la formule qui represente l'action mutuelle de deux portions infinitement petites de conducteur voltaiques", L'Academie Royale des Sciences, Paris, June 10, 1822.

1.8   J.C. Maxwell, *A treatise on electricity and magnetism*, Oxford University Press, Oxford, 1873.

1.9   F.E. Neumann, "Die mathematischen Gesetze der inducirten elektrischen Stroeme", Akademie der Wissenschaften, Berlin, 1845.

1.10   F.E. Neumann, "Ueber ein allgemeines Prinzip der mathematischen Theorie inducirter elektrischer Stroeme", Akademie der Wissenschaften, Berlin, 1847.

1.11   O.D. Kellog, *Foundations of potential theory*, Dover, New York, 1929.

1.12   P. Graneau, *Ampère-Neumann electrodynamics of metals*, 2nd Edition, Hadronic Press, Palm Harbor FL, 1994.

1.13   F.E. Neumann, *Vorlesungen ueber elektrische Stroeme*, Teubner, Leipzig, 1884.

1.14   H.G. Grassmann, "A new theory of electrodynamics", Poggendorf's Annalen, Vol.64, p.1, 1845.

1.15   H.A. Lorentz, *The theory of electrons*, Teubner, Leipzig, 1909.

1.16 G.T. Fechner, "About the link between Faraday's induction phenomena and Ampère's electrodynamic phenomena", Poggendorf's Annalen, Vol.64, p.337, 1845.

1.17 W. Weber, *Elektrodynamische Maasbestimmungen ueber ein allgemeines Grundgesetz der elektrischen Wirkung*, Wilhelm Weber's Werke, Springer, Berlin, 1893.

1.18 G. Kirchhoff, *Gesammelte Abhandlungen*, Ambrosius Barth, Leipzig, p.131, 1882.

1.19 A.K.T. Assis, "Deriving Ampère's law from Weber's law", Hadronic Journal, Vol.13, p.441, 1990.

1.20 G. Kirchhoff, "Ueber die Bewegung der Elektricitaet in Draehten", Poggendorf's Annalen, Vol.100, 1857.

1.21 G. Kirchhoff, "Ueber die Bewegung der Elektricitaet in Leitern", Poggendorf's Annalen, Vol.102, 1857.

1.22 P. Graneau, A.K.T. Assis, "Kirchhoff on the motion of electricity in conductors", Apeiron, No.19, p.19, 1994.

Experimental Demonstration of Longitudinal Ampère Forces

Ampère Tension

When two Ampèrian current elements lie on a straight line and point in the same direction, the angles of Ampère's force law are $\varepsilon=0$ and $\alpha=\beta=0$ or $180°$. This reduces the angle function of Eq.1.24 to -1 and the mutual force between the elements is then

$$\Delta F_{m,n} = i_m\, i_n\, \frac{dm\ dn}{r_{m,n}^2} \tag{2.1}$$

This latter expression is always positive and therefore represents repulsion. If the two elements belong to the same rigid metallic conductor they will create tension in the interatomic bonds between the elements. This will be called Ampère tension.

When dealing with a liquid conductor, the atomic bonds are absent and Eq.2.1 generates compression outside and beyond the element pair in question. This illustrates how the mechanical properties of the conductor affect the resulting longitudinal Ampère forces. Mechanical considerations determine the outcome of experiments as much as the electrodynamic force law. Critics of the Newtonian electrodynamics have tended to overlook this fact.

When the two co-linear elements belong to separate solid metal circuits, their interaction does not contribute to the generation of tension in either one of the circuits, but will strain the structure that keeps the circuits in place.

The order of magnitude of Ampère tension and tensile stress are indicated by the graphs of figure 2.1. Tension and stress depend quite strongly on the shape and size of the conductor cross-section and are greatest for round conductors. The plot of Ampère tension versus current indicates that the effect is almost negligible below 10 kA and very large above 100 kA. In mega-ampere circuits Ampère tension is likely to be the dominant design parameter.

Most conductors used for the transmission and distribution of electricity carry less than 10 kA continuously. The largest power conductors have cross-sectional areas from 50 to 100 cm^2 in order to keep them cool and waste as little energy as possible. Ampère tensile stresses in the widely used copper and aluminum conductors of the utility industry are therefore less than 1 N/cm^2. They are negligible compared to the weight-dependent stresses in hanging conductors. This explains why Ampère tension has gone unnoticed in a century of electric power distribution.

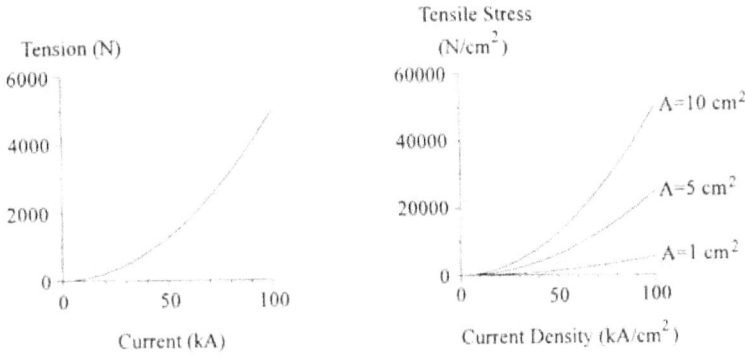

Figure 2.1 : Order of magnitude of Ampère tension and stress

Substantially larger currents may flow for brief periods of time when power circuits are accidentally short-circuited or struck by lightning. Fault currents of this nature are known to have reached 100 kA. Hence power conductors may occasionally experience tensile impulses up to 5000 N lasting for a few cycles of the power frequency. Since the Ampère tension is proportional to the square of the instantaneous current, AC current will set up tension corresponding to the root-mean-square amplitude pulsating unidirectionally at twice the AC frequency. It is possible that 100 kA fault current pulses have fractured conductor joints, but no proof is available.

Ampère tension plays a major role in pulse power circuits where currents in excess of 100 kA are commonplace. Examples are railguns and other electromagnetic accelerators, plasma fusion experiments, the simulation of the electromagnetic pulse (EMP) of a nuclear explosion, exploding wires and fuses, commutating opening switches for the discharge of inductively stored energy, pulse magnets, and so on.

Most of the Ampère tension is being generated between close-neighbour current elements. It is a localized phenomenon which occurs not only in straight wires but in all circuit shapes. Wherever an electric current flows in a solid metallic conductor, there will be tension induced in the atomic bonds of the lattice.

The question arises: can close-neighbour repulsion be cancelled by an opposing interaction with other elements in a more remote branch of the circuit? The available evidence suggests that the tension can never be eliminated and, if anything, may be enhanced by current in a return branch of the circuit. This conclusion was drawn from an analysis of a square circuit. Computer assisted finite current element analysis was used for the computations. The principle working rules of the finite current element analysis used here, are:

1.   Current elements are *volume* elements designed to fill the space occupied by the conductor metal.

2.   Experience has shown that the length-to-width ratio of the element must be approximately one, or the calculated forces will diverge widely from measurements.

3.   The location of the current element is a point at its geometrical center.

4.   Strings of touching elements are aligned along current streamlines. A single string of elements will be called a current filament.

5.   At circuit corners and everywhere along a curved filament, there may have to be some overlap of adjacent elements. This is the major defect of finite current element analysis.

An infinitely long conductor can be treated as a closed circuit, as it was in Grassmann's electrodynamics. Yet it would be futile to analyze it because the infinite number of elements will give infinite tension at every point along the conductor. To prove anything about Ampère tension, the investigation has to focus on closed metallic circuits of finite size. In a straight portion of a finite circuit the elemental repulsion of Eq.2.1 will create tension. In order to determine to what extent this tension is modified by the presence of the remainder of the circuit, we consider the example illustrated in figure 2.2. This shows a square circuit, carrying a steady current i, which is adequately cooled to ensure constant temperature. Sides BC, CD and AD are firmly embedded in a dielectric structure which is fixed to the laboratory frame.

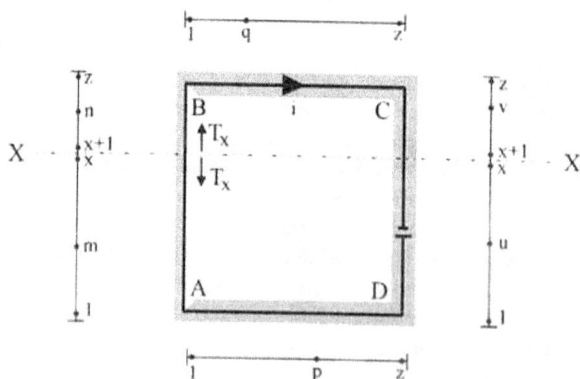

Figure 2.2 : Square circuit with one free side

AB is a free length of wire resting against a wall meant to absorb the lateral force on it.

Let T_x/i^2 be the specific tension in interatomic bonds across the plane XX intersecting the wire AB. As further shown by figure 2.2, each side of the square is assumed to be divided into z equal-length elements which are thin enough for the conductor to be treated as a single filament.

A major contribution to T_x comes from the repulsion of the general elements m in AX by the general elements n in XB. Since Eq.2.1 is independent of the unit of length, we may choose this to be

$$dm = dn = 1 \text{ unit of length} \tag{2.2}$$

With the labelling of current elements indicated on figure 2.2, the distance between the two general elements may be written

$$r_{m,n} = n - m \tag{2.3}$$

The dimensionless specific tension (tension/square-of-current) contribution by the m-n element combination is

$$\frac{T_1}{i^2} = \sum_{m=1}^{x} \sum_{n=x+1}^{z} \frac{1}{(n-m)^2} \tag{2.4}$$

This is a maximum when x=z/2.

Next we consider the interaction of current elements in AB with other elements in sides BC and AD. The interactions in question are all repulsions. This is due to the fact that the angle function $(2\cos\varepsilon - 3\cos\alpha \cos\beta)$ is negative for all relevant element combinations. Now some assumption must be made about the mechanical behaviour of the unsupported wire AB. It is very thin compared to its length and will, therefore, have little strength as a strut, while being quite strong in resisting tension. Hence it may be treated as an ideal string, recognizing that this must involve some approximation. Repulsions between BC and BX are taken up by the tensile strength of the wire BX and do not exert tension in the atomic bonds across plane XX. The same is true for interactions between AD and AX. The repulsions between BC and AX, as well as between AD and XB, do however add to T_x. This is due to AX and XB having no column strength, and their consequent yielding under axial compression is necessary for tension to be felt at the XX plane. By resolving the latter repulsions along AB we obtain the second contribution to the specific tension across plane XX, that is

$$\frac{T_2}{i^2} = \sum_{n=x+1}^{z} \sum_{p=1}^{z} \frac{3}{r_{p,n}^2} \cos^2\alpha_n \sin\alpha_n + \sum_{m=1}^{x+1} \sum_{q=1}^{z} \frac{3}{r_{q,m}^2} \cos^2\alpha_m \sin\alpha_m \qquad (2.5)$$

where

$$r_{p,n}^2 = (n-0.5)^2 + (p-0.5)^2 \qquad (2.6)$$

$$r_{q,m}^2 = (z-m+0.5)^2 + (q-0.5)^2 \qquad (2.7)$$

$$\cos\alpha_n = \frac{n-0.5}{r_{p,n}} \qquad , \qquad \sin\alpha_n = \frac{p-0.5}{r_{p,n}} \qquad (2.8)$$

$$\cos\alpha_m = \frac{z-m+0.5}{r_{q,m}} \qquad ; \qquad \sin\alpha_m = \frac{q-0.5}{r_{q,m}} \qquad (2.9)$$

The 0.5-terms arise from the fact that the position of the current element is a point halfway along its length.

The third contribution to T_x derives from interactions between AB and CD. The angle function for this pair of sides always has $\cos\varepsilon = -1$ and $-\cos\beta = \cos\alpha$. Furthermore, since α varies from $45°$ to $135°$, $2\cos\varepsilon - 3\cos\alpha\cos\beta = -2 + 3\cos^2\alpha$, is never positive. As a result of Eq.1.24, all interactions are again repulsions.

It is convenient to split CD by the plane XX with general elements u on one side and v on the other. Symmetry ensures that every elemental repulsion with an upward longitudinal component is offset by a symmetrical interaction with a corresponding downward component. Therefore actions of XC on XB do not contribute to Tx. The same is true for actions of DX on AX. Tensile forces will, however, be produced in AB by the actions of XC on AX and by DX on XB. They give

$$\frac{T_3}{i^2} = \sum_{m=1}^{x} \sum_{v=x+1}^{z} \frac{-1}{r_{m,v}} (-2 + 3\cos^2\alpha_v)\cos\alpha_v$$

$$+ \sum_{n=x+1}^{z} \sum_{u=1}^{x} \frac{-1}{r_{n,u}^2} (-2 + 3\cos^2\alpha_u)\cos\alpha_u$$

$$(2.10)$$

where

$$r_{m,v}^2 = (v-m)^2 + z^2 \qquad (2.11)$$

$$r_{n,u}^2 = (n-u)^2 + z^2 \qquad (2.12)$$

$$\cos\alpha_v = \frac{v-m}{r_{m,v}} \qquad (2.13)$$

$$\cos\alpha_u = \frac{n-u}{r_{n,u}} \qquad (2.14)$$

The total specific tension in the wire may then be obtained by adding Eqs.2.4, 2.5 and 2.10.

$$\frac{T_x}{i^2} = \frac{T_1}{i^2} + \frac{T_2}{i^2} + \frac{T_3}{i^2} \qquad (2.15)$$

Figure 2.3 is a plot of the three tension components and their sum for $z = 1000$. In the middle of AB the tension is seen to be largely due to repulsion of co-linear elements. Near the ends of AB it is mostly produced by actions across the corners A and B. Side CD makes only a small contribution to the tension in AB.

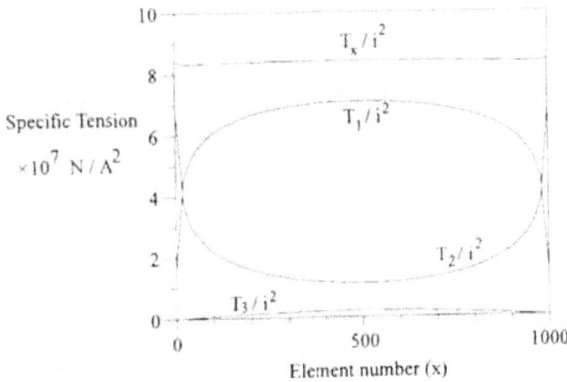

Figure 2.3 : Specific tension in free side of square circuit (figure 2.2)

The computed tension increases with z, the number of elements per side. The current element is a volume element and its width should be equal to its length. Hence for a single filament calculation, the appropriate z-value is determined by the conductor diameter and the length of the side of the square.

To show the variation of Ampère tension with conductor diameter, we let z vary from 20 (thick) to 200 (thin). On one end of this scale the width of the wire is 1/20th of the side of the square and on the other it is 1/200th of the same length. As a first step we calculate the greatest tension contribution, given by Eq.2.4, across the mid-plane of the square at $x = z/2$.

Table 2.1 lists the results. A regression analysis performed on this data revealed a very close fit to

$$\frac{T_1}{i^2} = 0.19 + \ln z \qquad\qquad (2.16)$$

z	T_1/i^2
20	3.188
50	4.103
100	4.796
150	5.202
200	5.489

Table 2.1 : Computer evaluation of Eq 2.4 for z varying from 20 to 200 and $x = z/2$.

For $z = 1000$, Eq.2.16 gives the specific tension of 7.098 compared to 7.099 obtained by finite element analysis. This very good agreement instills confidence in extrapolations of Eq.2.16 to much larger values of z which, otherwise, would have to be obtained by excessive computing expenditure. It can be shown that the specific tension contributions T_2/i^2 and T_3/i^2 obey similar logarithmic laws. Hence T_x/i^2 will also be a logarithmic function of z.

Eq.2.16 however tends to infinity with z. This is reasonable because the number of element interactions also tends to infinity. If the current element is assumed to be infinitely divisible, the Ampère electrodynamics becomes absurd, resulting in infinite forces. Therefore there is no alternative but to accept that elements have finite size. This is, in any case, a requirement of the atomicity of matter. Could the limiting size of the current element be the distance between neighbouring atoms of the metal lattice? This would be of the order of 10^{-7} cm. It would amount to 10^9 current elements in AB of figure 2.2, if the latter side is 100 cm long. Eq.2.16 then gives a specific tension of 20.91, which is only three times the tension obtained for $z = 1000$. It is not an unreasonably large number and lends support to the idea that the atomic cell, or the atom itself, is the extent of the basic current element.

By the rules of finite current element analysis, the specific tension of 20.91 would apply to a conductor diameter of 10^{-7} cm. For conductors of larger diameter, and maintaining the atomic current element concept, the conductor has to be resolved into a bundle of parallel filaments, each being a string of atoms. How the bunching of filaments leads to the reduction in Ampère tension will be discussed in Chapter 3. This effect of tension reduction with increasing conductor diameter has been called 'longitudinal force dilution'.

Wire Rupture by Current Pulses

Ampère reviewed his electrodynamics research most completely in reference [2.1] which was re-published as recently as 1958. Even though there is indirect evidence for the existence of longitudinal forces in a number of his experiments, his force law was simply a generalization of many experimental results obtained with copper wires and liquid mercury cups and troughs. Blondel [2.2] in her excellent historical review of Ampère's achievements points out that the Frenchman in June 1822 claimed that his formula required that co-linear current elements should repel each other, and that this prediction was going to be put to a test by A. de la Rive of Geneva. The experiment was performed shortly thereafter with Ampère present in de la Rive's laboratory. It will be referred to as Ampère's 'hairpin experiment'. At the time it was considered to have been an unqualified triumph of Ampère's theory. Subsequent to the formulation of the Grassmann-Lorentz force law there arose much controversy about the hairpin experiment. This will be fully discussed later in this chapter. Faraday repeated the hairpin experiment in London, and Ampère wrote to him [2.2] in 1825 that, in some way, this experiment revealed the fundamental fact of electrodynamics.

Until about 1960 all experimental evidence for longitudinal Ampère forces had been collected with circuits containing some liquid metal which was usually mercury. This permitted the detection and measurement of relatively small forces with steady currents of 1000 A or less. The analysis of electrodynamic phenomena in liquid metals is quite complex. It is usually referred to as magneto-hydro-dynamics (MHD). Ampère made no allowance for MHD effects. This contributed to the controversy surrounding the hairpin experiment.

Early in the 1960's, Nasilowski [2.3-2.5] in Warsaw performed decisive experiments without liquid metal which revealed the existence of Ampère tension. He was studying the behaviour of copper fuse wires when subjected to a sudden current pulse. The pulse amplitudes were quite small, up to approximately 2000 A, but the pulse duration was relatively long, of the order of 50 ms. Nasilowski's current pulses were supplied by throwing a short-circuit across the output of a rotating generator. He found that with stepwise increases in current amplitude, in successive tests, a situation would be reached when a straight wire would fracture in the solid state at one or more places along its length. As a result of the stored inductive energy in the circuit, an electric arc in air formed immediately across each fracture gap and the current continued to flow without interruption. Figure 2.4 shows open-shutter photographs of three wire disintegrations of 0.5 mm diameter copper wires of 56 cm length at three different current amplitudes. The smallest current produced the arcs of (a) and the largest the arcs of (c).

To prove the existence of Ampère tension in wires conclusively, it has to be shown that: (1) The current pulse produces impact-tension fractures, across grain boundaries, without neck-down. (2) The fracture faces must be approximately perpendicular to the wire axis. It has sometimes been argued that the wire fractures could be the result of Coulomb explosions due to the aggregation of positive or negative charges. Coulomb forces however are multi-directional and would split the wire lengthwise as well as transverse, and in other directions. (3) The tension could not be the result of thermal shock due to the sudden heating of the wire.

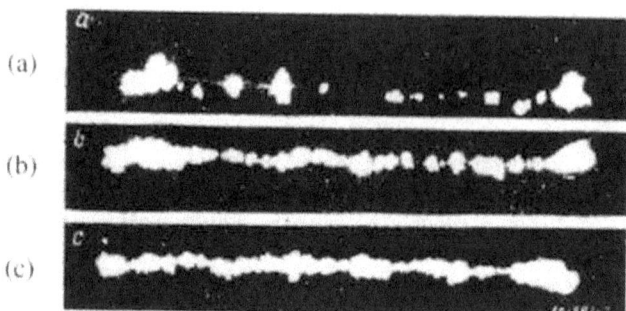

(a)

(b)

(c)

Figure 2.4 · Wire fracture arcs (Nasilowski [2.3])

Nasilowski was not aware of Ampère's force law, and undertook careful metallurgical examinations of the wire fragments in search of an explanation of the unexpected phenomena. In the process he established the first two conditions (brittle fracture and transverse fracture faces) of the proof of the existence of Ampère tension.

Sectioning wire fragments longitudinally, Nasilowski [2.4] found some incipient transverse cracks which would have lead to clean breaks had the current lasted longer. Microscopic evidence of the grain structure on the fracture faces and the cracks proved conclusively that atomic bonds had ruptured in tension before any melting had taken place. But the air arcs bridging the gaps subsequent to rupture of the lattice produced micron-thick layers of molten material, clearly recognizable by their dendritic nature as post-fracture melting on top of the metal grain structure.

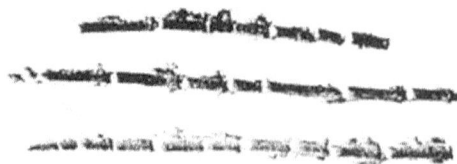

Figure 2.5 · Copper wire fragments (Nasilowski [2.4])

Figure 2.5 shows a collection of Nasilowski's wire fragments which fell on a metal tray below the horizontally strung test wire. The transverse nature of the fractures is clearly visible. This rules out Coulomb-type explosions. The fragments of figure 2.5 belong to a 1 mm diameter copper wire. Oxidation of copper and the arc melting produced the fuzz on the outside of the fragments. The oxidation is absent when aluminum wires are exploded.

The authors repeated Nasilowski's experiment in modified form at the Massachusetts Institute of Technology in 1982 [2.6, 2.7]. Aluminum wires of 99 percent purity and 1.2 mm diameter were subjected to current pulses of 5000 to 7000 A amplitude. The current was derived from a high-voltage capacitor bank and passed through an inductor to allow it to ring down at 2 kHz over a period of five to ten milli-seconds. When the capacitor bank was charged to 60 kV, the discharge current would decay approximately exponentially, as shown by the oscillogram of figure 2.6(a), without breaking the wire. This was accompanied by a wire temperature rise of several hundred degrees centigrade and probably caused a thermal expansion of the order of one percent.

By subsequent increases of the discharge voltage in steps of 2 kV, a pulse current level was reached at which the wire broke in one or more places. When repeating the test with a new wire and 2 kV additional voltage, the wire would break into a greater number of pieces. At the 66 and 68 kV levels, a one meter long wire would fragment into 20 - 30 pieces. Finally, at 70 kV the test wire would show signs of melting which obliterated much of the tensile break evidence. The 68 kV oscillogram of figure 2.6(b) indicates quenching of the discharge current due to the many arcs across fracture gaps.

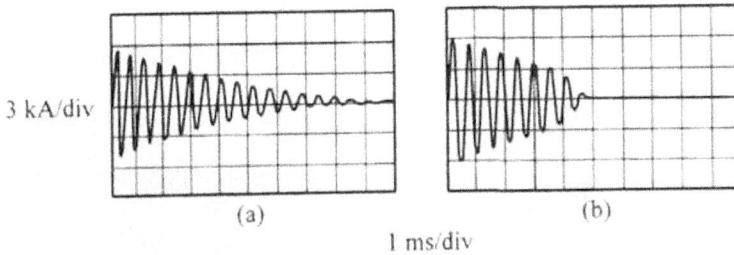

Figure 2.6 : Discharge current oscillograms, (a) at 60 kV, (b) at 68 kV

The MIT experiments were performed with the test wires hung vertically with cotton threads, leaving 1 cm long arc gaps in air between the wire ends and the two terminations of the capacitor-inductor circuit, as shown in figure 2.7. When the discharge switch was closed, the two air gaps would break down, allowing the current pulse to flow through the wire. The purpose of the arc gaps was to allow distortion free thermal expansion of the wire and mechanical decoupling from the rest of the circuit.

Figure 2.8(a) shows a collection of aluminum wire fragments produced by these experiments. Photograph (b) depicts transverse fractures which were spot-welded together again by arcing across the fracture gaps.

Figure 2.8(c) is an optical micrograph of one end of a wire fragment. It illustrates the transverse brittle fracture across grain boundaries. Photograph (d) of figure 2.8 was taken with a scanning electron microscope. Similar micrographs of greater magnification revealed the micron-deep melting of the fracture surface due to arcing across the fracture gap.

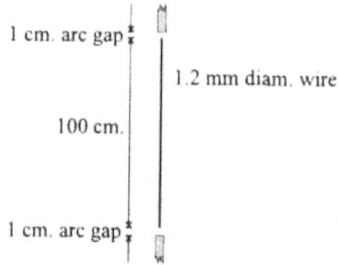

Figure 2.7 : Suspension of wire to be fragmented in MIT experiment

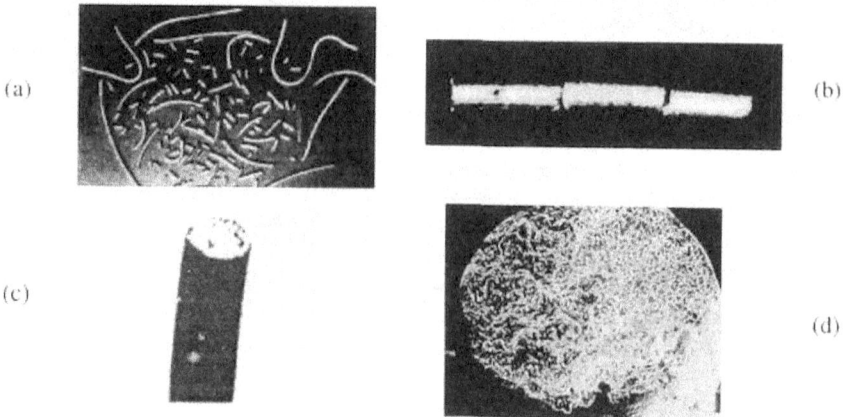

(a)

(b)

(c)

(d)

Fragments of a 1.2 mm diameter aluminium wire
(a) collection of fragments from several wire explosions
(b) fragments reconnected by arc spot welding
(c) optical micrograph of fracture face
(d) scanning electron micrograph of fracture face

Figure 2.8 . Photographs of wire fragments from MIT experiment

When the wire is treated as a bundle of current filaments, the transverse pinch forces may be calculated with Ampère's or the Lorentz force law. Both laws predict the same pinch forces. They are capable of extruding soft wire. Northrup [2.8] calculated the longitudinal thrust which would be created in a liquid metal column which was subject to the normal electrodynamic pinch forces. His calculation revealed a value of this thrust, which turns out to be about ten percent of the Ampère tension which would be set up by the same current in a solid conductor. No reduction in wire diameter due to pinch forces has been observed in the experiments just described. This is not surprising in view of Northrup's result. There appears to be no connection between pinch forces and the brittle fractures observed in wire explosions.

Another possibility, which has been considered, is that pinch forces cause elastic stress in the longitudinal direction, and that the wire fractures as a result of travelling and reflected stress waves. The velocity of sound in aluminum is of the order of 5000 m/s. Hence a stress wave could travel the length of the wires used by Nasilowski and the authors in 0.1 to 0.3 ms. Tensile stress magnification by multiple reflections is, therefore, not out of the question.

The elastic stress waves would have had to be generated by oscillations of the pinch forces. Such oscillations existed in the MIT experiments where the current frequency was 2000 Hz. But Nasilowski used a unidirectional (dc) current pulse, and the pinch force was not removed from the wire until the current ceased to flow. Only after the end of the current pulse could pinch force relaxation have produced travelling stress waves. Nasilowski proved, however, that the wire ruptured into many pieces well before the end of his pulse. This has eliminated elastic stress waves as the cause of wire ruptures.

The following quantitative considerations support the Ampère tension mechanism. A 100 cm long, 1.2 mm diameter aluminum wire used in the MIT experiments weighed three grams. For a peak current amplitude of 6000 A, it should have experienced a calculated maximum Ampèrian tensile stress of 2264 N/cm². This is equal to the ultimate strength of the material at around 300°C. The impact strength of the metal will be less. Therefore the first break in the wire could have occurred quite early in the capacitor discharge cycle, provided the fracture pieces managed to separate sufficiently before the current dropped to a level at which there was not enough force. The repulsion of two adjacent fragments is equal to the Ampère tension just before the break. If the break occurs halfway along the wire, calculations have shown that the wire portions should have been subject to an acceleration of 1713 times that of gravity. In 0.1 ms this would have produced a wire separation of 0.08 mm. This seems adequate for a clean break.

The MIT experiments furnished two more pieces of experimental evidence supporting the Ampère tension process. The first one concerned the location of the first wire rupture. One function of the arc gaps was to decouple the wire from the remainder of the circuit so that the specific tension would vary along the length of the wire as seen with T_1/i^2 of figure 2.3. In other words the tension would be a maximum in the middle of the wire. That is roughly where the first break was found to occur when the current was small enough to produce only one break.

The MIT experiments also revealed that by careful adjustment of the discharge current, the one meter long wire could be shattered into hundreds of solid pieces of unequal lengths. After the first rupture, the two resulting pieces of wire were found to break near their midpoints, and so the subdivision process continued until current flow stopped or a certain minimum length of wire fragment was reached. Experience has shown that this minimum length, below which fragments refuse to subdivide, is approximately one wire diameter. Nasilowski [2.4] discovered the same irreducible fragment length with copper wires. He further proved that this minimum length would persist in the metal vapour pattern, if the current pulse was powerful enough to fully evaporate the wire. (see figure 7.3)

The reason for the minimum fragment length remained a mystery for nearly thirty years until it was realized that Ampère's force law itself contains the explanation. This was the second of the additional experimental facts which confirmed the existence of Ampère tension.

To calculate the minimum fragment size consider the square-section conductor shown in figure 2.9. This is subdivided into $5 \times 5 = 25$ square filaments. Each filament consists of cubic elements carrying the current i. Ampère tension forces exist across any section S of the conductor. This tension will be proportional to i^2. The total conductor current is 25i.

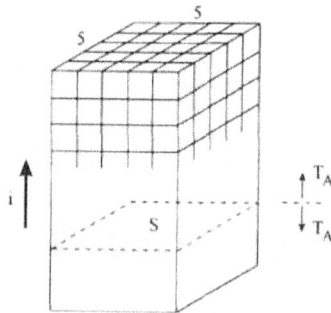

Figure 2.9 : Square conductor column divided into 25 square current filaments

According to the rules of Newtonian stress analysis, the Ampère tension can be calculated with Eq.1.24 by summing the force components $\Delta F_{m,n}$ of element pairs having one member on either side of S. It is convenient to define the positions of the elements of a pair by (u, v, m) and (x, y, n) as shown in figure 2.10.

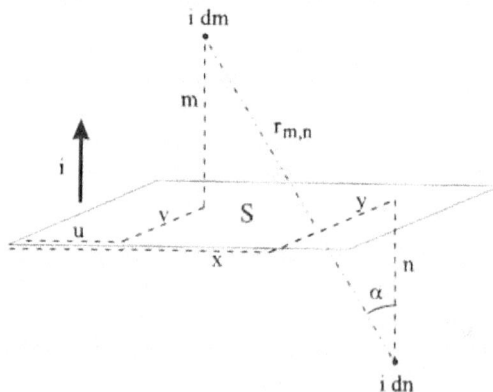

Figure 2.10 : Coordinates of two current elements on opposite sides of the transverse plane S of figure 2.9

As before, we choose the unit of length to be the edge of one cubic current element, which is then equal to dm and dn. For the parallel current elements of figure 2.9 we have $\varepsilon = 0$ and $\alpha = \beta$, so that Ampère's law (in MKS-A units) may be written

$$\frac{\Delta F_{m,n}}{\frac{\mu_0}{4\pi} i^2} = \frac{2 - 3 \cos^2\alpha}{r_{m,n}^2} \tag{2.17}$$

The left-hand side of this equation has the dimension

$$\frac{N \; m}{H \; A^2} = \text{dimensionless} \tag{2.18}$$

On the right-hand side of Eq.2.17 we have $r_{m,n}$ which is the number of elements which can be fitted into the distance $r_{m,n}$. Hence Eq.2.17 is a dimensionless equation. The force described by it is a 'specific force' and will be denoted by $\Delta f_{m,n}$, such that

$$\Delta f_{m,n} = \frac{\Delta F_{m,n}}{\frac{\mu_0}{4\pi} i^2} = \text{dimensionless} \tag{2.19}$$

Similarly, there exists a specific tension t which is a dimensionless normalized form of the actual tension T, so that for the Ampère tension T_A we may write

$$t_A = \frac{T_A}{\frac{\mu_0}{4\pi} i^2} = \text{dimensionless} \tag{2.20}$$

In the MKS-A system of units specific forces have to be multiplied by $(\mu_0/4\pi) i^2$ to convert them to actual forces.

With the coordinates of figure 2.10 we have

$$r_{m,n} = \sqrt{(m+n-1)^2 + (u-x)^2 + (v-y)^2} \tag{2.21}$$

$$\cos\alpha = \frac{m+n-1}{r_{m,n}} \tag{2.22}$$

Eqs.2.17, 2.21 and 2.22 enable us to calculate the specific force $\Delta f_{m,n}$ for any current element pair with one member of the pair above and the other below the plane S across which T_A is being determined. Next the specific forces have to be resolved in the vertical direction (multiplied by $\cos\alpha$) and summed over the six coordinates to obtain the specific tension

$$t_A = \sum_v \sum_u \sum_x \sum_y \sum_m \sum_n \Delta f_{m,n} \cos \alpha \qquad (2.23)$$

Surprising results emerge from this analysis. What would have been a repulsion between two thin conductor portions in the single-filament approximation may change to an attraction in the multi-filament model. The deciding factor in every elemental force is the sign of the numerator of Eq.2.17. The change-over from repulsion to attraction occurs when

$$2 - 3 \cos^2\alpha = 0 \qquad \text{or} \qquad \alpha = 35.3° \qquad (2.24)$$

This is the 'magic angle' to which Grassmann objected in 1845 on non-scientific grounds (see figure 1.15). Greater values of α give rise to attraction. The mixing of repulsions and attractions in a contiguous group of elements can reduce the Ampère tension to zero or negative values. For example, the interaction force between the two element layers on either side of the interface S, in figure 2.9, would be an attraction which adds to the binding forces of the lattice. If we add a layer on each side of S, the net force across S would still be attraction. But with the addition of further layers T_A goes to zero. Beyond this number of layers the Ampère tension becomes positive and then there exists a tendency for the wire segment to rupture into two pieces.

In more accurate finite element calculations, estimates were made of the minimum fragment length for which the Ampère tension across the mid-plane comes to zero. A square $28 \times 28 = 784$ filament model was used instead of the $5 \times 5 = 25$ filaments shown in figure 2.9. The mid-plane tension came to zero for a fragment length which was 1.4 times the width of the square conductor. This indicates that there is minimum fragment length in the wire rupturing process. The agreement of this calculation with experimental observations is remarkable and further reinforces the proof of the existence of longitudinal Ampère forces.

In the 170 years since Ampère announced his empirical force law, not a single experiment with metallic conductors has come to light which has contradicted his law. This is an outstanding record, particularly because the law disagrees with the Lorentz force law of modern electromagnetism. It may be possible to formulate an alternative explanation for any one longitudinal force experiment, but as convincing as this alternative may be, it cannot overturn Ampère's law in its many other experimental confirmations.

Since the publication of the Ampère force explanation of wire ruptures by current pulses [2.6], in 1983, only two investigators have published their speculations with regard to an alternative explanation. No new experimental evidence was offered. Aspden [2.9, 2.10] argued that the lattice fractures may be caused by induced electromotive forces. He agreed with Neumann, Faraday, and Maxwell that induced e.m.f's act on the electric fluid and not on the metal atoms. The quantum mechanical treatment of conduction in metals has replaced the older electric fluid model by the Fermi sea of conduction electrons. It allows virtually no mechanical coupling between the conduction electrons and the lattice ions. If it were otherwise, conductors would be dragged along by the electric current. No evidence whatsoever has been found for such drag forces. Hence forces exerted on the conduction electrons cannot possibly break the strong atomic bonds between adjacent lattice ions. This represents ample grounds for refuting the induction hypothesis of wire ruptures.

Another factor denies the electromotive fracture mechanism. Assuming that conduction electrons are bonded to the lattice by some obscure field interaction, it still does not allow atomic bonds to be broken by induction forces because induction is a one-way action. Without a reaction force on the source electron, the lattice cannot be strained even with a bond between the conduction electron and an associated lattice ion. One-way inductive interactions have been discussed in Chapter 1 in conjunction with Neumann's laws of induction. In this respect the Faraday-Maxwell induction hypothesis does not differ from Neumann's.

The second alternative explanation of the wire rupture mechanism was advanced by Ternan [2.11] in 1987. He maintained that, provided the ends of the wire were free to recede from each other, as they were in the MIT experiments, rapid heating of the wire would generate a longitudinal expansion velocity. When the maximum temperature was reached, the wire ends would want to travel on because of inertia and this would create tensile stress. In Ternan's opinion this could result in a standing stress wave. According to his calculations the stress should have been sufficient to break the hot wire.

Ternan arrived at a stress wave velocity of 5.1 km/s which was 36 times the average thermal expansion velocity. If his analysis had been correct, the wires in Nasilowski's experiments [2.4] should not have been fractured because their ends were clamped to the laboratory frame, and not free to move as required by Ternan's explanation. Nasilowski's wires however, broke into as many pieces as were produced in the MIT series of wire explosions.

The authors performed a specific experiment [2.12] to disprove the Ternan proposal. For this purpose the wire was threaded through a glass tube and its ends were clamped to the laboratory frame. If Ternan had been correct, the wire should not have fractured and it should have become straight again, after cooling down. Instead the wire was found curled up in the glass tube and broken. It had thermally expanded, coiled up in the tube, broke in the expanded state, and the resulting fragments contracted, leaving them in the distorted form in which they were found.

Electromagnetic Jets in Mercury Channels

The second longitudinal force experiment to be described has become known as the 'straight-through mercury channel experiment'. It was inspired by the research of Carl Hering (1860-1926) who discovered the electromagnetic pinch effect. The idea of the simple experiment [2.7] which we performed at MIT is depicted in figure 2.11.

Two ½-inch-square copper bars were glued into a close fitting long rectangular groove, machined into a plastic board. The 30 cm long length between the copper bars was filled with liquid mercury until the liquid level was flush with the top of the horizontal copper bars. In this way the cross-section of the mercury conductor was made the same as that of the copper conductors, except for the meniscus deformation on top of the mercury surface, due to the surface tension of the liquid metal. The experiments were carried out with steady DC currents flowing along the straight copper-mercury-copper conductor. Currents up to 1000 A were supplied by a controlled DC power source. Current flow times were severely restricted, sometimes to a few seconds, to ensure that the temperature rise in the mercury was not more than 50°C.

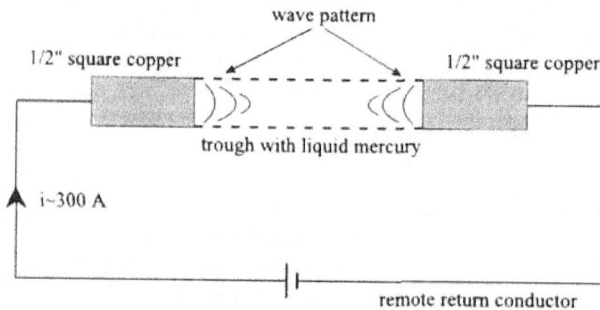

Figure 2.11 : Straight-through liquid mercury channel

When about 300 A was flowing along the channel, a wave pattern became apparent on the free liquid mercury surface. The waves disappeared almost instantaneously when the current was switched off, showing that they were not caused by the temperature of the liquid. The waves were strongest at the two copper-mercury interfaces, and they died out within a few centimeters.

Conventional theory claims that the only significant electrodynamic forces on the mercury section are the forces which pinch the liquid equally strongly all along its length. No pinch deformation could be seen anywhere on the mercury surface. This led to the conclusion that the wave disturbances were caused by longitudinal Ampère forces.

Looking at one of the square copper interfaces, finite element computations with Ampère's force law revealed that the mercury should have been more strongly repulsed from the copper near the center of the square section than at the corners and the periphery of the conductor. Hence if the longitudinal forces gave rise to hydrostatic pressure in the liquid, this should have resulted in mercury flow away from the copper along the conductor axis. The flow must then return to the copper against the weaker longitudinal forces near the conductor periphery. The wave motion confirmed the return flow just below the surface tension skin on the upper mercury surface.

Internally generated forces in the liquid cannot interfere with this flow pattern. To see this more clearly, consider an isolated pair of current elements in the liquid. They may repel or even attract each other and move apart or closer together. This motion will not change the position of the center of mass of the element pair relative to the laboratory. Hence this force pair cannot contribute to the jet motion which sweeps a body of liquid along the channel.

When electric current flows from a solid into a liquid conductor, or vice versa, the longitudinal Ampère forces always produce a jet of liquid which streams away from the interface. The jet propulsion forces arise between those element pairs which have one member in the solid conductor and the other member in the liquid conductor. Every electromagnetic jet so produced has a reaction force in the solid conductor. This will be proved with other experiments in which the solid conductor is mobile in a bath of liquid mercury.

The wave motion in the straight-through channel proved the axial flow in liquid mercury. Only the Ampère-MHD can explain how the jets are created. It furnishes further evidence for the existence of longitudinal electrodynamic forces.

When the current was gradually increased, the wave motion would become more violent and extended farther away from the interfaces. A point was reached, near 1000 A, when an arc was struck at one end of the mercury trough or the other. The arc was formed when all of the mercury had been pushed away from the copper conductor. This would reduce the current, allowing the mercury to flow back and close the gap. If the current was maintained, another arc would strike, and so on. There was no indication of the mercury level being depressed by pinch forces prior to arcing. The only explanation of this phenomenon, first discovered by Hering, was that the longitudinal Ampère forces become so strong that they separate the liquid from the solid conductor. It was, in fact, the same mechanism by which current pulses ruptured wires.

The collision of jets in the mercury channel, emanating from the two solid electrodes, should become stronger, the shorter the trough section. With a very short trough of 0.4 cm length, it was found that 1000 A of DC current would suddenly expel virtually all the liquid mercury upward into the air. Prior to this explosion, the liquid metal was seen to bulge up, as if a bubble was forming underneath, and then collapse again. In order to escape from the channel, the mercury had to pull a vacuum underneath itself. Electrodynamic forces competing with atmospheric pressure caused the bulging and subsequent collapse of the liquid metal.

Ampère's Hairpin Experiment

Ampère himself appears to have been under some pressure to explicitly demonstrate longitudinal electrodynamic forces. He had taken the view that his empirical law was the generalization of many experimental results, collected mainly by himself, which all implied the existence of longitudinal forces and no explicit demonstration was needed. Yet at the invitation of the Swiss scientist de la Rive, both men performed a famous test in Geneva in 1822. This will be called the 'hairpin experiment' to distinguish it from the many other demonstrations. The experiment was actually designed by de la Rive. Ampère's sketch of the apparatus is reproduced in figure 2.12.

ABCD is a circular dish filled with liquid mercury. The liquid metal is divided into two pools by the insulation barrier AC. Current leads m and s dip into the two mercury pools. A current source has to be connected between the terminals E and F. In Ampère's time this was a battery of galvanic cells. An insulated copper wire with bare ends r and n, in the shape of a hairpin, floats on the mercury with the two parallel legs np and qr straddling the insulation barrier. The copper wire bridge pq passes over the barrier. When the terminals E and F are connected to the galvanic cells, current will flow from m across a short distance of mercury to n, then mainly along the copper wire from n to p over the bridge to q and back to r. The current is then returned through another short distance in mercury to the terminal sE.

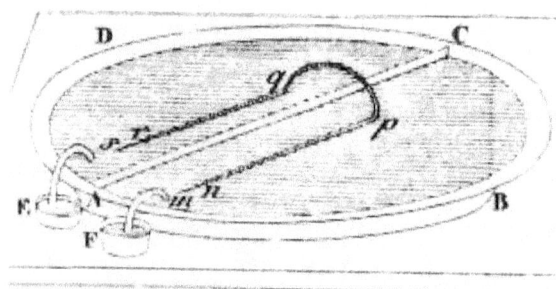

Figure 2.12 : Ampère's sketch of his 'hairpin' experiment

Ampère and de la Rive observed that the current would make the hairpin float away from the terminals toward C, as expected. They considered this to be proof that longitudinal reaction forces existed between current elements external to the copper hairpin, and current elements in np and qr. Ampère no doubt realized that, according to his own law, a very small transverse repulsion force should have been exerted by the battery circuit on the hairpin bridge, pq. Given the large distance between the mercury current elements and the bridge, this force was considered to be negligible compared to the longitudinal repulsion forces. The hairpin legs were made longer than the bridge to ensure that the transverse force on the bridge was very small. In 1822 the Geneva experiment was considered to be an unqualified success, demonstrating the existence of longitudinal electrodynamic forces.

Figure 2.13 shows a diagram of the circuit with which the authors [2.15] in 1981 performed the hairpin experiment at MIT. The hairpin cdefg was an insulated copper conductor with bare end-faces at c and g floating on two liquid mercury channels ab and a'b'. When more than 200 A of current was passed through this circuit, the hairpin moved to the ends of the channels b and b'. According to Ampère's law, most of the motive force was provided by the repulsion between the hairpin ends and the mercury with which the ends were in contact.

A new observation was made at MIT which was not reported by Ampère and de la Rive, nor by anyone else of the many scientists who repeated the original experiment over the years. When the forward motion of the hairpin was blocked by an obstacle, strong jets of liquid mercury could be seen to emanate from the hairpin ends c and g. The turbulence in the liquid gave the distinct impression of the hairpin being subject to jet propulsion. The area of strongest turbulence was quite narrowly defined to the hairpin ends. The jet effect becomes unmistakable at 500 A and so strong at 1000 A that there is danger of liquid mercury splashing out of the troughs. In 1822 Ampère and de la Rive had no instruments to measure current, and their currents were probably too weak to clearly show the jets.

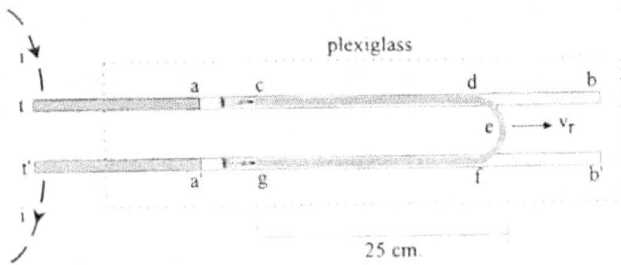

Figure 2.13 MIT version of Ampère's 'hairpin' experiment

At MIT it was also noticed that similar jet turbulence occurred at a and a'. This observation, together with Hering's published experience, led to the design of the previously described straight-through mercury channel experiment shown in figure 2.11. The overall turbulence in the liquid mercury sections ac and a'g could be increased, at constant current, by moving the hairpin closer to the copper bars. All of these facts are consistent with the formation of electrodynamic jets. The reaction forces of the two jets at the ends of the hairpin legs are obviously the propulsion forces of the hairpin motion away from the current source. Ampère and de la Rive, therefore, were correct when they claimed that the hairpin experiment verified the existence of longitudinal forces.

Ampère's critics of recent times have held that the motive force on the hairpin is the Lorentz force on the bend e of figure 2.13, passing over the dielectric barrier. The following argument by Hillas [2.16] is typical. The magnetic field at the bend is primarily due to the current in the hairpin legs. Therefore one might expect that the reaction force should reside in the hairpin legs. This is not the case because the Lorentz force on the legs is perpendicular to the hairpin motion. In any case, as will be discussed in the next section, the special theory of relativity requires the Lorentz force on the hairpin bend to be generated by energy-momentum impact. This makes the reaction to the motive force, or the magnetic pressure, a force on the field, that is on vacuum!

It turns out that the magnitude of the Lorentz force on the hairpin bend is equal to the sum of the two Ampère jet reaction forces. Now since there is no Lorentz force which could produce the jets, and no Ampère force which accounts for the transverse force on the hairpin bend, one of the two theories must be wrong and the other will be right. The observation of the electromagnetic jets in the mercury decides this issue in favour of Ampère's electrodynamics. Once more the existence of longitudinal electrodynamic forces is confirmed. The relevant experiment has been performed 173 years ago.

Hillas [2.16] acknowledged the existence of the mercury jets, but attributed no reaction forces to them. That this absence of a reaction force flagrantly violated Newton's third law did not concern him. Special relativity is well known to disagree with Newtonian mechanics.

The Electrodynamic Impulse Pendulum

The electrodynamic impulse pendulum is a large-scale version of Ampère's hairpin experiment, suitable for momentum measurements. It was invented by Pappas [2.17] and first published in 1983. The purpose of the Pappas experiment, performed at the University of Athens in Greece, was to show that the Lorentz force could not be generated by field energy impact.

The authors repeated Pappas' experiment at MIT. Figure 2.14 shows the setup of the MIT electrodynamic impulse pendulum. The 'hairpin' was made of a copper strip 0.5 inch high and 0.05 inch thick. The strip formed two one-meter long sides and one 30 cm short side of an open rectangle. The pendulum was suspended from the ceiling by four 2.56 m long cotton threads. The horizontal displacement of the pendulum was measured with the cardboard slide C resting lightly on a flat table top.

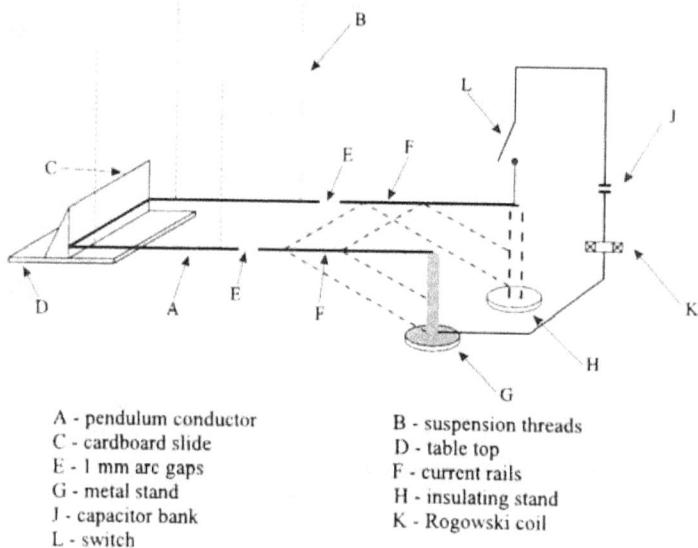

A - pendulum conductor	B - suspension threads
C - cardboard slide	D - table top
E - 1 mm arc gaps	F - current rails
G - metal stand	H - insulating stand
J - capacitor bank	K - Rogowski coil
L - switch	

Figure 2.14 : Electrodynamic impulse pendulum used at MIT

Current pulses through the pendulum circuit were derived from an 8 μF high-voltage capacitor bank which could withstand voltage reversals up to ±100 kV. The capacitor discharge was initiated by dropping a mechanical switch arm L. Two parallel current rails F, of the same copper strip of which the pendulum was made, brought the current to the hairpin via two, one millimetre long, arc gaps in air, E. The rails were supported on two heavy stands and carefully aligned with the horizontal legs of the hairpin pendulum.

To perform a momentum experiment, the capacitor bank was charged to a voltage between 30 and 80 kV. The switch was then dropped, causing arcing between current rails and pendulum legs. The damped oscillatory current pulse was recorded with the Rogowski coil K, and an oscilloscope. The current pulse did cause the pendulum to swing away from the current rails and move the cardboard slide through a distance s, subsequently measured with a ruler. The duration of the current pulse was a fraction of one milli-second, while almost all of the pendulum displacement occurred after the current had ceased to flow.

If u is the initial horizontal velocity of the pendulum, its momentum, mu, imparted to it by the current pulse, may be calculated from the pendulum string length R, the pendulum mass m, and the cardboard slide displacement s. Energy conservation gives

$$m\,g\,h \;=\; \frac{1}{2} m\,u^2 \qquad \text{or} \qquad u \;=\; \sqrt{2\,g\,h} \qquad\qquad (2.25)$$

where g is the acceleration due to gravity, u the maximum horizontal velocity that would be attained in the limit when the impulse duration tends to zero, and h is the maximum vertical lift of the pendulum. Figure 2.15, relates the height h to s, R and the pendulum angular displacement, θ.

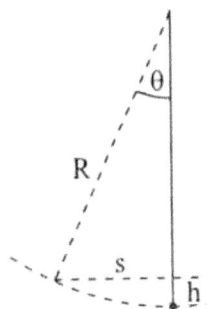

Figure 2.15 : Pendulum parameters

h may be derived from the two simultaneous equations

$$h \;=\; R\,(1 - \cos\theta) \qquad\qquad (2.26)$$

$$s \;=\; R\,\sin\theta \qquad\qquad (2.27)$$

leaving

$$h = R \left(1 - \sqrt{1 - \left(\frac{s}{R}\right)^2} \right) \approx \frac{s^2}{2R} \qquad (2.28)$$

This approximation assumes s<<R, and is accurate to three significant figures and may be used in Eq.2.25 to calculate u.

This experiment is one of many in which, numerically, the Ampère and Lorentz laws predict roughly equal force magnitudes. However this particular arrangement can be used to demonstrate that the two laws do not agree on where and how the forces are applied to the circuit. In MKS-A units, the electrodynamic impulse imparted to the pendulum by the Lorentz force $F_L = 10^{-7} k i^2$ may be written

$$P_1 = 10^{-7} k \int_0^\infty i^2 \, dt \qquad (2.29)$$

where i is the instantaneous current and k (N/A²), the specific force, must be determined from the geometry of the circuit. The capacitor discharge current i decayed approximately exponentially and could be written

$$i = e^{(t/T)} I_0 \sin \omega t \qquad (2.30)$$

where T was the time constant with which the oscillation decayed. The full amplitude I_0 would have been reached if the circuit had contained negligible resistance. It was possible to integrate Eq.2.30, as required by Eq.2.29, to obtain

$$P_1 = 10^{-7} k I_0^2 \left[\frac{T}{4} - \left(\frac{\frac{1}{T}}{\left(\frac{2}{T}\right)^2 + (2\omega)^2} \right) \right] \qquad (2.31)$$

with $\omega = 2\pi f$ being the radian angular frequency. As far as the pendulum experiments were concerned, the second term of Eq.2.31 was negligible. Hence the electrodynamic impulse could be taken to be

$$P_i = 10^{-7} k I_0^2 \left(\frac{T}{4} \right) \qquad (2.32)$$

The magnitudes of I_0 and T were measured on the pulse current oscillogram, and the constant k was evaluated by finite current element analysis. For the analysis the conductor strip was subdivided into ten parallel square-section filaments, and each filament was further subdivided into cubic current elements. On the assumption of uniform current distribution, that is with the same current flowing in each filament, the result was $k=9.27$ (N/A²). At the observed ringing frequency of 15.7 kHz, the current distribution must have been non-uniform with strong concentrations at the strip edges. When all the current was divided between the two edge filaments, and no current was assumed to flow between them, the result came to $k=11.0$ (N/A²). In the evaluation of the experiment the smaller figure $k=9.27$ (N/A²) was used.

Without further thought, it might be expected that the calculated electrodynamic impulse P_i of Eq.2.32 should be equal to the initial pendulum momentum mu, where the mass m of the pendulum was weighed and the initial velocity u was determined from the pendulum swing and Eqs.2.25 and 2.28.

The first observation was that the pendulum would not swing properly in the forward direction, but tended to turn in the horizontal plane, unless the hairpin legs were perfectly aligned with the current rails. This immediately suggested that the pendulum was being pushed from behind by the longitudinal Ampère forces on the ends of the hairpin legs. A slight misalignment would then be expected to convert a true forward push to an unstable turning moment. If the propulsion force had been the Lorentz force pulling at the front of the pendulum, any misalignment would have corrected itself without noticeable turning. The turning behaviour, therefore, provided immediate evidence in favour of longitudinal Ampère forces.

With good alignment the pendulum swing was in the forward direction and the cardboard slide moved forward with it. The magnitude of the initial momentum, determined from the cardboard slide and the pendulum mass, was, however, only a fraction of the electrodynamic impulse P_i of Eq.2.32. Initially this fraction was of the order of twenty percent. Both Ampère's law and the Lorentz law predict the same total propulsion force. Therefore the low observed momentum could not be attributed to one law giving a smaller force than the other. The explanation had to reside in the mechanical response of the pendulum to forces located in different parts of the hairpin.

Another problem with the experiment was the flutter of the pendulum legs and current rails. Much of this was probably caused by the transverse Ampère or Lorentz forces on the parallel conductors. The flutter was eliminated by using insulating struts to reinforce the pendulum and the rails by strong cross-bracing. This increased the weight of the pendulum and led to a smooth forward swing.

A consequence of the reinforcement of the conductor structure was a large increase in the measured pendulum momentum for a given current pulse. The transverse forces on the parallel conductors A and F of figure 2.14 could not be contributing to longitudinal momentum generation. Thus it was realized that the mechanical reinforcement of the hairpin pendulum also prevented a certain amount of buckling of the pendulum legs, in response to the longitudinal Ampère forces. It was this effect of the reduced buckling which seemed to give rise to the momentum increase.

The largest calculated impulse force which was applied to the pendulum was 3300 N. A quick force of this magnitude on the ends of the hairpin legs is likely to distort and buckle the weak copper strip. The reinforcement of the hairpin made it stiffer and more resistant to buckling, thus allowing more of the impulse force to generate momentum.

The Lorentz law predicts that the pendulum is pulled from the front, rather than pushed from the rear. The pulling action would be incapable of producing conductor buckling or any significant pendulum deformation. Only the longitudinal Ampère forces can account for the momentum increase due to the stiffening of the pendulum structure.

Table 2.2 lists the results obtained with the largest current pulse of 60 kA amplitude. The efficiency of the pendulum as a momentum generator is defined by mu/P_i. This came to 0.76. Before the reinforcement of the pendulum structure, the efficiency was only of the order of 0.2. The step-up in efficiency proves that the pendulum was pushed by longitudinal Ampère forces, rather than pulled by Lorentz forces.

Measured

Current:	ringing frequency	$f = 15.7 \, kHz$
	time constant	$T = 0.27 \, ms$
	maximum amplitude	$I_0 = 60 \, kA$
Pendulum:	mass	$m = 0.815 \, kg$
	displacement	$s = 10.95 \, cm$
Energy stored in capacitors		$U_e = 25.6 \, kJ$

Calculated

Pendulum:	initial velocity	$u = 21.44 \, cm/s$
	initial momentum	$mu = 0.1747 \, kg \, m/s$
Max. Lorentz and Ampère force		$F = 333.7 \, N$
Electrodynamic impulse		$P_i = 0.2253 \, N \, s$
Efficiency		$mu/P_i = 0.76$

Table 2.2 : Pendulum Results

Pappas [2.17] arrived at the same conclusion by a different route. He demonstrated that there was insufficient energy in the field to produce the Lorentz force by energy-momentum impact. This left the longitudinal Ampère forces as the only viable alternative which could generate the measured momentum of the electrodynamic impulse pendulum.

Ever since the elastic properties of the ether were abandoned to accommodate the special theory of relativity, it was felt that another mechanism for absorbing reaction forces in the field (vacuum) was required. With the impulse pendulum, the relativistic [2.20] reaction force has to oppose the Lorentz force exerted on the pendulum front end. In other words the rate of change of electromagnetic momentum in the field must support the force which accelerates and decelerates magnetic energy to and from the velocity of light c. The energy momentum density is related to the Poynting vector by

$$\vec{p} = \frac{1}{c^2} (\vec{E} \times \vec{H}) \qquad (2.33)$$

where $\vec{E}$ and $\vec{H}$ are the electric and magnetic field strengths at a point. The volume integral of the rate of change of this momentum density, over all space, gives the vacuum reaction force

$$\vec{F}_{vac} = \int \frac{d\vec{p}}{dt} \, dv = \frac{1}{c^2} \int \frac{d}{dt} (\vec{E} \times \vec{H}) \, dv \qquad (2.34)$$

where t stands for time and v for volume. When the integral is not taken over all space, then the rate of change of momentum may be smaller than indicated by Eq.2.34. It is customary to make up the difference by the surface integral of Maxwell's stress tensor over the finite volume of the momentum integral. In the context of the present investigation the integral will be taken over infinite space to avoid complications with Maxwell stresses which have physical meaning only in the old ether theory.

For any instant in time, the vacuum reaction force may be written

$$\vec{F}_{vac} = \frac{d}{dt} (m_e c) \qquad (2.35)$$

where m_e is the equivalent electromagnetic mass of the magnetic energy stored in the field. Since c is constant, the vacuum force will exist only during changes of m_e, that is when magnetic energy is emitted into the field, or absorbed from the field, by a conducting body. The amount of field energy U_f that must at any time be associated with the vacuum reaction force is

$$U_f = m_e c^2 \qquad (2.36)$$

This is the famous mass-energy relation of special relativity.

If F_{vac} is at all times the simultaneous reaction force to the Lorentz F_L, when the latter accelerates a metallic conductor of real mass m from zero to the velocity u, then momentum conservation requires

$$m u = \int p \, dv = m_e c \qquad (2.37)$$

The field energy which has to be emitted or absorbed by the conductor to comply with Eq.2.37 is

$$U_f = m_e c^2 = m u c \qquad (2.38)$$

When evaluating Eq.2.38 for the experimental data listed in table 2.2, that is for mu=0.1747 kg.m/s, the required field energy is found to be 52.4 MJ. Table 2.2 also shows that the energy stored in the capacitors and expended during the impulse was 25.6 kJ. This means 2000 times the energy available would have been required to generate the Lorentz force by energy-momentum impact. This constituted Pappas' proof that the electrodynamic impulse pendulum could not possibly be driven by the Lorentz force. Consequently, the longitudinal Ampère forces must exist.

Ampère Tension or Hoop Tension in Wire Circles?

As previously pointed out, the Ampère tension mechanism does not only arise in straight wire sections. It is also operative in curved conductors because most of the tension is due to very local repulsion between adjacent and almost adjacent current elements. If the Ampère electrodynamics is applied to a circular current loop, every tangential atomic bond will be found to be subject to tension. In reference [1.12] it was argued that the Ampère tension is superimposed on the hoop tension, the latter being due to the radially outward directed transverse forces on the current elements. In view of the better understanding of the Newtonian electrodynamics existing today, this claim has to be re-examined. For this purpose consider a circular turn carrying an instantaneous current i.

The magnetic pressure of field theory is exerted on the inside of the wire loop and tends to move every current element, of the single-filament model, radially outward. Were it not for the restraining forces of the metal lattice, the circle would expand. The restraining forces are believed to be the tangential hoop tension forces. A similar mechanism is operative in a circular pressure vessel where the magnetic pressure is replaced by gas pressure.

To calculate the hoop tension, T_H, by the finite current element technique (single-filament approximation), we must first compute the Lorentz force on one element due to the magnetic field of all other elements. This is equivalent to using Grassmann's law (see Chapter 1) for all current element interactions. Let this elemental Lorentz force be ΔF_L. Because of symmetry, ΔF_L will be the same for all elements of the circle. The elemental forces act simultaneously and radially outward. The hoop tension can be found by calculating the reaction force ($2T_H$, because there are two break points) between any two semicircles into which the wire loop can be divided. The mutual reaction force is easily computed by resolving the ΔF_L forces perpendicularly to the bisecting diameter and summing the resolved components over one semicircle. The T_H value so obtained agrees with experiment.

This computational process allows self-forces to contribute to the mutual reaction force between the two semicircles, into which the current circle has been divided. To see this more clearly, take two current elements i.dm and i.dn on the same semicircle. Element i.dm contributes to $F_{L,n}$ at dn, and element i.dn contributes to $F_{L,m}$ at dm. In other words, part of the mutual reaction force $2T_H$ between the two semicircles is actually the result of self-interactions of current elements in the same rigid conductor portion. If the internal self-interactions are omitted, T_H will be reduced and then no longer agree with measurements. The self-interactions clearly violate Newton's third law! However, in the special theory of relativity, of which the Lorentz force is a vital part, the Newtonian mechanics is invalid, and it is in fact necessary to violate the third law in order to achieve compliance with experiment.

How does the Ampère electrodynamics generate the tension T_A in the wire circle? Again we must calculate the mutual reaction forces between any two semicircles, but this time in compliance with Newton's third law, that is without internal self-interactions in one semicircle. Every current element pair which contributes to the reaction force $2T_A$ must be split between the two semicircles. This is an essential requirement of Newtonian stress analysis, which in the American Institute of Physics Handbook [2.21] is defined as:

"A stress is a force per unit area with which the part of the medium on one side of an imaginary surface acts on the part of the other side."

The imaginary surface is the perpendicular plane which bisects the current circle.

Within the errors of the single-filament approximation it has been found that $T_A - T_H$. It is believed that when the wire is subdivided into many parallel filaments, the tensions would be identical ($T_A = T_H$). That the two force laws give the same total reaction force between two parts of the same circuit has already been discussed in connection with the hairpin experiment. Differences between the two laws, however, can exist in the force distributions.

In the Ampère tension calculations it is found that most of T_A is caused by adjacent and near-neighbour elements on either side of the stress plane. The remote elements make only a small contribution to the tension. In the relativistic electrodynamics it is the elements furthest away from the stress plane which contribute most to the reaction force. Unfortunately this difference cannot be put to an experimental test.

The analysis does suggest another question. Is the tension in the wire circle really hoop tension, as predicted by the magnetic pressure concept and the Lorentz force law? If the answer turns out to be 'no', then longitudinal Ampère forces would be confirmed once more. This question was investigated at MIT [2.22].

An 8 µF high-voltage capacitor bank was discharged through a 1000 µH inductor to produce exponentially decaying oscillatory current pulses of 10 kA amplitude and ringing down at 2 kHz over a period of 5 to 10 ms. The discharge currents were passed through a 1.2 mm diameter aluminum wire bent into the shape of a semicircle. As shown in figure 2.16, the wire semicircle of 25 cm radius was suspended with insulation filaments in a vertical plane, leaving 1 cm long arc gaps to connect it to the terminals of the discharge circuit. The arcs allowed distortion free thermal expansion. More importantly, since the arcs have no tensile strength, no hoop tension could be produced anywhere in the semicircle. Hence if the Lorentz force mechanism applies, current pulses could not produce tensile fractures of the semicircle.

The following test procedure was adopted. The capacitor bank was charged to 50 kV and then discharged through the circuit containing the wire semicircle by closing the mechanical switch of figure 2.16. This would heat the aluminum wire but did not break it. The discharge current oscillogram was recorded. It revealed the maximum current amplitude and exponential damping. After the wire had cooled down to room temperature, the experiment was repeated with 52 kV, and subsequently at 2 kV increments. At 62 kV and a peak current of approximately 6000 A, the wire would fracture in two or three pieces. A certain amount of melting could always be found at the arc gap ends but none could be seen at the fracture faces.

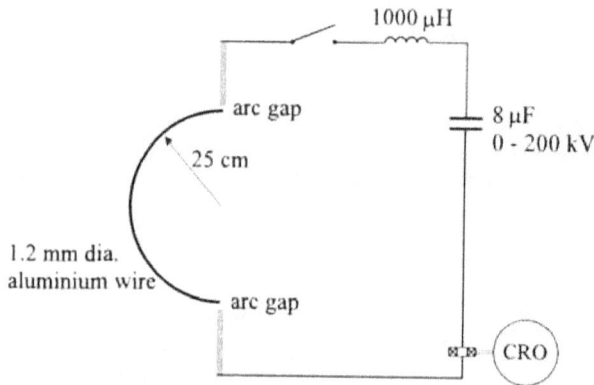

Figure 2.16 : Semi-circular wire fragmentation experiment

After the first breaks a new semicircular wire was mounted in place and the discharge experiment was repeated with a further 2 kV voltage increment. This produced a greater number of wire fragments, just as in the case of straight wires (see figure 2.7). The appearance of the fragments was similar to those shown in figure 2.8. No hoop tension could possibly exist in the wire semicircle of figure 2.16. The Lorentz forces on the semicircle should have deformed the wire and accelerated it sideways. They could not have produced the observed solid state fractures. At one and the same time it had been established, therefore, that the relativistic model of hoop tension formation with self-forces is wrong, and the Ampère tension model is correct.

It is instructive to examine how Ampère tension is distributed in the semicircle, because it is this distribution which gives rise to the multitude of fractures. To do this we have to determine the reaction forces between any two portions of the semicircle. This gives the tension across the dividing surface.

For this purpose the single-filament semicircle AXB of figure 2.17 is divided into z equal elements of arc, each subtending an angle of

$$\Delta \theta = \frac{\pi}{z} \qquad (2.39)$$

The elements along XA are labelled 1, 2, ... m, ...x; and those along XB are labelled 1, 2, ... n, ...(z-x). The distance between two general elements m and n is denoted by $r_{m,n}$, and the arc mXn subtends the angle $\theta_{m,n}$.

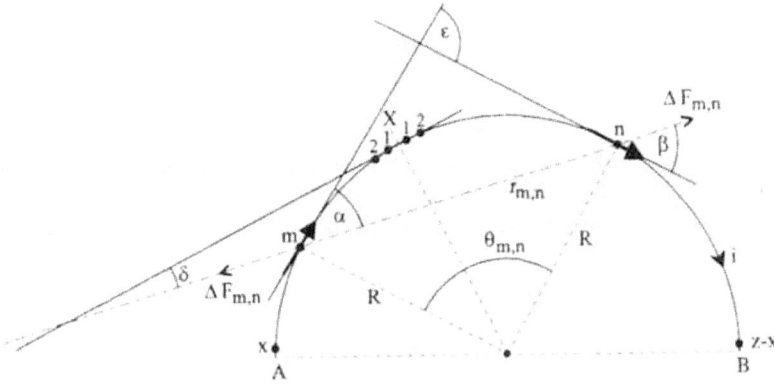

Figure 2.17 : Construction for tension calculation in semi-circle

In this case the angles of Ampère's law, Eq.1.24, obey the relationships

$$\alpha = \beta \qquad (2.40)$$

$$\varepsilon = \theta_{m,n} = 2\,\alpha = 2\,\beta \qquad (2.41)$$

If R is the radius of the semicircle, then

$$dm = dn = \frac{\pi}{z}\,R \qquad (2.42)$$

and

$$r_{m,n}^2 = 2\,R^2\,(1 - \cos\theta_{m,n}) \qquad (2.43)$$

It can easily be shown that for any element combination on the semicircle, the angle function of Eq.1.24 is negative and therefore all the Ampèrian interactions are repulsions. With Eqs.2.39 and 2.43 Ampère's force law may be written

$$\frac{\Delta F_{m,n}}{i^2} = -\left(\frac{\pi}{z}\right)^2 \frac{2 \cos \varepsilon - 3 \cos^2 (\varepsilon/2)}{2 - 2 \cos \varepsilon} \tag{2.44}$$

where

$$\varepsilon = \left(\frac{\pi}{z}\right)(m + n - 1)$$

This repulsive force can be resolved into components which are tangential and perpendicular to the semicircle at the point X. The perpendicular components will tend to move the semi-circle along the line OX, and the tangential components create tension at X. To find this component we must multiply the elemental force by $\cos(\delta)$, where δ can be seen in figure 2.17, and is equal to

$$\delta = \left(\frac{\pi}{z}\right)\left(\frac{n - m + 1}{2}\right)$$

Thus the elemental tension can be written

$$\frac{\Delta T_{m,n,X}}{i^2} = \frac{\Delta F_{m,n}}{i^2} \cos(\delta) \tag{2.45}$$

Therefore the wire tension across the perpendicular plane OX is given by

$$\frac{T_x}{i^2} = \sum_{m=1}^{x} \sum_{n=1}^{z-x} \frac{\Delta F_{m,n}}{i^2} \cos(\delta) \tag{2.46}$$

For a semicircle of 1000 elements and varying values of x, the computer evaluation of Eq.2.46 gave the results plotted on figure 2.18.

The graph shows that the specific Ampère tension T_x/i^2 does not vary greatly except at the ends of the semicircle. This tension variation is in fact similar to that in a straight wire which is mechanically disconnected from the rest of the circuit. Hence a current pulse which is able to rupture a straight wire will also fragment a semicircle. This deduction from finite current element calculations with Ampère's force law has been fully confirmed by the experiment of figure 2.16.

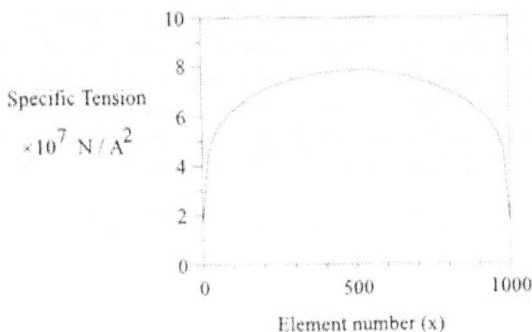

Figure 2.18 Computed specific Ampère tension in semi-circle

In a series of four papers, Christodoulides [2.23] argued that the Lorentz and Ampère force laws were identical in their predictions about the outcome of any experiment. Despite this sweeping claim, Christodoulides did not demonstrate how the Lorentz force could explain any of the experiments described in this chapter and in a number of previous publications. In fact nobody has been able to furnish these explanations. It has to be concluded, therefore, that the Christodoulides mathematics is insufficient to prove the equivalence of the two laws.

Christodoulides calculated the reaction forces between a single current element and the remainder of the closed metallic circuit, first with the Lorentz force law and then with Ampère's law. As has been known since Neumann's time, in this particular case the two laws are in full agreement with each other. Christodoulides then assumed that these single element forces are the only forces which have to be taken into account when computing the interaction forces between two macroscopic portions of the metallic circuit. In relativistic field theory there is no other choice. As has already been shown several times, this procedure gives the correct total reaction force, but results in a distribution of this force which disagrees with experiment. Christodoulides never calculated the distribution of the Ampère reaction forces, and therefore missed the important difference between the relativistic and Newtonian force laws.

Neumann's Longitudinal Force Experiment

The existence of longitudinal Ampère forces was fully accepted during most of the nineteenth century. Neumann demonstrated them routinely to his students with a classroom experiment. Figure 2.19 is a diagram of his demonstration as recorded by one of his pupils [1.13].

A, B, and C are mercury troughs and D and E were copper wire bridges from A to B and from B to C. When current was passed along the troughs, the two pieces of wire separated further from each other. They appear to have been subject to longitudinal repulsion.

This experiment has sometimes been criticized because of the small hooks at the ends of the wires which dip into the liquid metal. Transverse forces on these short sections could cause the longitudinal wire repulsion. To clarify this issue, the authors analyzed the circuit of figure 2.19 with finite elements. The result of this analysis was published in reference [1.12]. The Ampère repulsion was found to be more than hundred times as large as the transverse forces on the hooks. Besides, the absolute value of the transverse forces was too small to even overcome the strong adhesion of copper to liquid mercury.

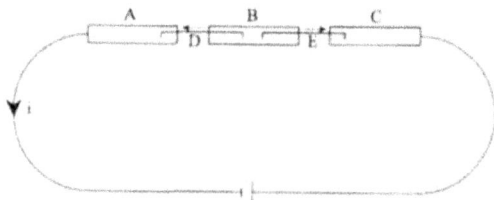

Figure 2.19 : Neumann's longitudinal force demonstration

To eliminate the hooks on the wire bridges, the authors devised a modified version of Neumann's test for longitudinal forces. The apparatus involved the straight-through mercury channel of 30 cm length depicted in figure 2.11. The cross-section of the copper-mercury-copper conductor was 1.27×1.27 cm^2. The circuit was closed by a remote return conductor through a 500 A DC current supply. Two insulated copper rods AB and CD, as shown in figure 2.20, of 5 cm length and 0.3 cm diameter, with bare end-faces, were laid end-to-end on the mercury surface in the middle of the trough. Copper floats on mercury.

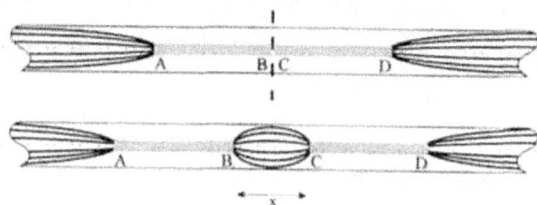

Figure 2.20 : Rod positions before and after passage of current

When a DC current of 450 A was switched to flow through the trough, the rods would submerge and separate axially. As soon as the current was switched off, ten or twenty seconds later, the rods would surface and were found to be separated by the distance x shown in figure 2.20. On account of the 50:1 resistivity ratio of liquid mercury to solid copper, the rods carried a substantial fraction of the total current in their section of the trough. As copper and mercury carried parallel currents, there existed lateral attraction between the two metals. These same lateral forces also cause the pinch effect. They urge each copper rod toward the center of the trough cross-section and thereby cause them to submerge. When the rods re-surfaced, they had separated by approximately 2.5 cm. Consider now the longitudinal

Ampère forces which must have caused the rod separation.

Figure 2.21 is a plot of the specific Ampère repulsion force (F/i^2) acting between the rods as a function of the distance of separation x shown in figure 2.20. The points on this graph were calculated with finite current element analysis in which the rods were replaced by single filaments of elements, the element length being equal to the rod diameter. According to this graph, the rods should strongly repel each other while they are in contact, and the repulsion force should fall off quite sharply with distance of separation. For 450 A the maximum repulsion comes to 0.075 N. This decreases to 0.01 N at x = 2.5 cm, which could well be the adhesion drag resisting further motion of the rods through the liquid mercury.

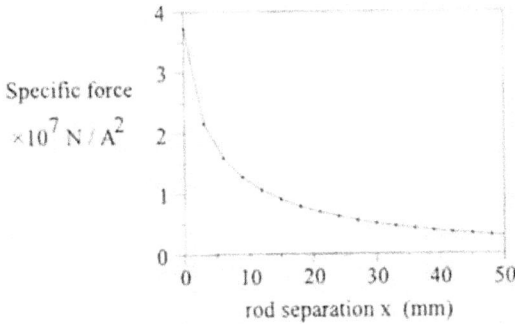

Figure 2.21 : Specific rod separation force as a function of x

There will be jets of mercury streaming away from both ends of each rod, as found in Ampère's hairpin experiment. These jets are associated with longitudinal reaction forces in the solid rods. When the experiment of figure 2.20 was performed with a single rod, this would simply submerge and re-surface at the same place. The jet forces at both ends were obviously of the same strength.

For short rod separations there will be a higher current density in the mercury between the rods than at their other ends. The Ampère electrodynamics predicts that, because of the higher density, the jets between the rods are stronger than the jets on the outside of the rods. Therefore the jet reactions reinforce rod-to-rod repulsion during the separation process.

Could transverse Lorentz forces have been responsible for the rod separation? Their action on the rods would have had no longitudinal component. Pinch forces on the mercury can, however, exert a longitudinal thrust on solid interfaces. On submersion, the pinch forces on the outer rod ends could have exerted longitudinal forces, albeit in the wrong direction. On the other hand, the pinch forces between the rods act in the direction of rod separation. On the whole they are an order of magnitude weaker than longitudinal Ampère forces.

Anyhow, in the beginning when the two rods were virtually in contact, with only a thin film of mercury between them, another effect, first observed by Roper [2.24], overpowers the pinch. Roper found that a thin mercury film bridging a gap in 2.5 mm diameter copper wire would be blown out of the gap, rather than pinched inward, with a current of 180 A.

The same phenomenon has been observed to occur in short arc gaps where the plasma is blown radially outward against pinch forces. It is a most important physical mechanism of electrodynamic arc explosions and filament fusion processes which are the subject of the last two chapters of this book. It will there be shown that the outward directed forces are caused by oppositely directed, that is colliding, electrodynamic jets. The jet reaction forces will indeed help to separate the rods. This mechanism renders the Lorentz pinch forces ineffective.

An explanation of the rod separation by thermal causes is ruled out because the temperature in the mercury bath was kept below 50°C. In any case, thermal effects would necessarily have been time cumulative, and would have revealed themselves by making the rod separation dependent on the duration of current flow. No such dependence was observed.

Without Lorentz force and thermal explanations of the rod separation of figure 2.20, experiment once more is in agreement with the existence of longitudinal Ampère forces.

The Liquid Mercury Fountain

In many ways the liquid mercury fountain of figure 2.22 [2.25] is the most beautiful of the longitudinal force experiments. It was originally performed in the authors' MIT laboratory. An insulated copper rod, bare on the end-face, projected through the bottom of a dielectric cup which was filled with liquid mercury. A copper ring electrode was partially submerged in the top surface. The cup was 4.5 cm deep and 6.4 cm in diameter. When 500 to 1000 A of DC current was made to flow between the rod and ring electrodes, a conical fountain head appeared on the free surface of the liquid metal, directly above the copper rod. Mercury could be seen to stream down the outer surface of the cone. While the current was sustained, the mercury flow and circulation was continuous.

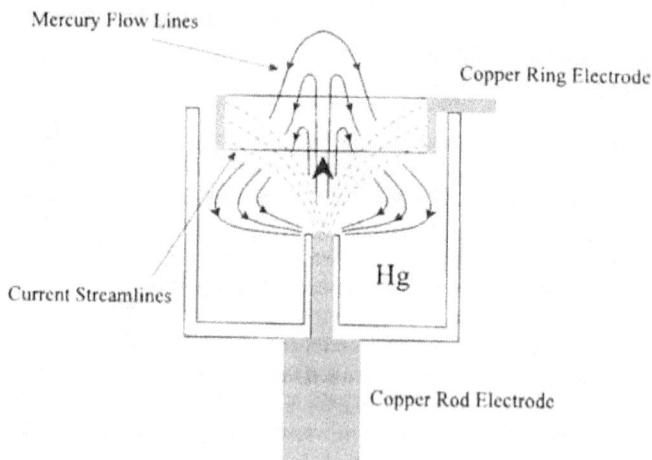

Figure 2.22 : Liquid mercury fountain

The dotted lines which are drawn in figure 2.22 from the rod electrode to the ring electrode represent current streamlines. The Lorentz forces are transverse to the streamlines and act radially inward and up on the current elements. In accordance with Newtonian mechanics, in order for this force to create the raised mound of mercury, there must be an equal and opposite thrust on another area. This must be the region below and outside the current streamlines. However, the continuous mercury circulation requires a flow of mercury across the streamlines, which cannot occur if these reaction forces exist. Only longitudinal Ampère force interactions between current elements in the copper rod and others in the mercury can explain the fountain jet without back-pressure in the liquid, which would stop mercury flow.

The existence of longitudinal Ampère forces in the liquid mercury fountain is also confirmed by the following quantitative considerations. When 1000 A of current were flowing, the fountain head was a cone of approximately 2 cm base diameter and 2 cm height. With the density of liquid mercury being 13.54 gm/cm^3, the head weighed 28.4 gm which corresponds to a force of 0.29 N. In any Newtonian theory there must exist an equal and opposite reaction force in the cup. Ampère's force law places this reaction force in the upper parts of the copper rod. This reaction, as well as the lift force, is given by

$$F_{lift} = \frac{\mu_0}{4\pi} k \ i^2 \quad (N) \qquad (2.47)$$

where the current i is measured in amperes. For the lift force of 0.29 N and a current of 1000 A, it is found that k=2.9. This is a typical value of the k-factor in jet propulsion examples. It is entirely consistent with the Ampère electrodynamics.

The pinch pressure in the liquid mercury would be a maximum if current flow was purely vertical and of the same diameter as the copper rod. Then from Northrup's theory [2.8] the axial thrust would come to

$$F_{pinch \ thrust} = \frac{\mu_0}{4\pi} \frac{i^2}{2} \quad (N) \qquad (2.48)$$

This is an upper limit of the pinch thrust and it could lift no more than 5 g of mercury above the free surface. The pinch effect, therefore, would not only stop mercury circulation, it would also be quantitatively incapable of lifting the fountain head.

Longitudinal Armature Motion

Mobile parts of metallic circuits are often called armatures. An example is the conductor bridge between the rails of a railgun. Contact is maintained by sliding friction, brushes, or arcs in air. Robson and Sethian [2.26], of the Naval Research Laboratory in Washington, performed an experiment with a vertically oriented tubular armature which, they hoped, would reveal longitudinal Ampère forces. This experiment is also described in Chapter 5, and the circuit is shown schematically in figure 5.8.

The 12.7 cm long copper armature was hung vertically from a rubber cord. The bottom of the armature made contact with the rest of the circuit by sliding through a telescopic tube joint, and the upper end passed through a hole in a horizontal aluminum plate with which the armature was in sliding contact. The remainder of the circuit consisted of a bottom plate and 12 conductor rods arranged around the sliding armature maintaining cylindrical symmetry. The rest of the circuit was comprised of a 65 μF capacitor which could be charged to 10 kV, and a spark-gap switch. In one particular test the capacitor was charged to 7.9 kV and the maximum discharge current through the armature was greater than 100 kA. The investigators expected longitudinal Ampère forces would lift the armature, resulting in easily observable up and down oscillations of it on the rubber cord. No lift and no oscillations of the armature took place.

Robson and Sethian had misunderstood the conditions required to observe longitudinal electrodynamic forces. They proved by their own calculations with Ampère's law that the longitudinal force on their armature was zero! In spite of this lengthy calculation, the authors drew the puzzling conclusion:

> "We have performed an experiment that should have provided a direct measurement of the longitudinal electrodynamic force, if such a force exists. No force was found."

In order to report something new, Robson and Sethian went on to claim that Ampère tension does not exist. They found the enthusiastic support of the editor of the American Journal of Physics (published mainly for conservative physics teachers) who was only too pleased to discredit the Newtonian electrodynamics. This editor subsequently rejected, without peer review, all submissions that enumerated the errors in the Robson and Sethian paper. The ink was hardly dry, when the same editor of AJP hailed the challenge to Ampère's law a Memorable Paper [2.27].

One of the errors made in the AJP paper was that Robson and Sethian subjected the Newtonian electrodynamic force law to the rules of relativistic mechanics. They overlooked the fact that tension is a Newtonian stress and has to be evaluated like any other Newtonian stress [2.28]. A definition of stress in a solid body is provided by Slater and Frank [2.29]. It reads:

> "To specify such a (stress) force, we imagine a surface element da to be drawn somewhere in the body, with a normal n. The material on either side of da exerts a force on the material on the other side; thus this force is a push normal to the surface if there is a pressure in the body, it is a tension if that is the form of stress, or it may be a shearing force, tangential to the surface."

An important corollary of this definition of Newtonian stress is that the interaction of two elements of matter on the same side of the stress surface makes no contribution to the stress at this surface.

Newtonian stress analysis appears to be no longer in the physics curriculum, but the subject is fully covered in engineering textbooks. The stress is felt by the atomic bonds which intersect the imaginary stress surface. The atoms themselves are not torn apart or compressed. Therefore, calculations of forces on atoms will not reveal stress. What has to be calculated is the force between atoms. This is to say, stress is the result of Newtonian action and reaction forces bridging the stress surface.

When stress is internally generated by mutual interaction forces between atoms, rather than by applied external forces on the body as a whole, we have to specify the interaction force by a formula which complies with Newton's third law. The two important electromagnetic forces which fulfill this condition are Coulomb and Ampère forces.

Consider first an example which involves Coulomb's law. This concerns a dielectric string charged along its length with additional electrons. Two electrons of charge e and distance $r_{m,n}$ repel each other with the force

$$\Delta F_c = k \frac{e^2}{r_{m,n}^2} \qquad (2.49)$$

It is immediately obvious that the string will find itself in tension everywhere except at the ends.

From the Newtonian definition of stress, the tension T at some surface which intersects the string is

$$T = k e^2 \sum_{m=1}^{x} \sum_{n=1}^{y} \frac{1}{r_{m,n}^2} \qquad (2.50)$$

where the electrons on one side of the stress surface are labelled 1, 2, ..., m, ..., x; and on the other side they are labelled 1, 2, ..., n, ..., y. Eq.2.50 does not give the sum of the force densities, that is forces on electrons, but rather the tension between electrons. A charged string would lie still and stretched on the laboratory bench.

A thin wire carrying a steady DC current will behave like the charged dielectric string if it is subject to Ampère's force law. That is to say, all the current elements (conducting atoms) in the wire would then repel each other. This is the mechanism which generates Ampère tension.

If the Lorentz force law is applied to this wire, it will not predict tension, because the Lorentz force must always be transverse to the wire axis. In the wire example we have found an instance in which the two force laws do make opposite predictions. A more dramatic difference between the laws can hardly be expected. Wire explosions have resolved this issue in favour of Ampère's law.

Critics of the wire example correctly argued that a complete circuit should have been considered, because forces due to the remaining parts of the circuit may cancel the tension. The square circuit of figure 2.2, examined earlier in this chapter did show some modification of Ampère tension due to the remainder of the circuit, but the tension was not cancelled.

We will now demonstrate the same fact with another example. This involves the rectangular circuit ABCD of figure 2.23. This circuit is assumed to carry a steady current i and stands in a vertical plane. X-Y is a horizontal surface which cuts the circuit in two parts, which are then electrically re-connected by thin liquid mercury films. The tension 2T (T from each leg) in this surface is equal to the upward force F_E on XBCY and the downward force F_E on YDAX. F_E is an experimentally determined force which has been measured for various rectangular circuits [1.12].

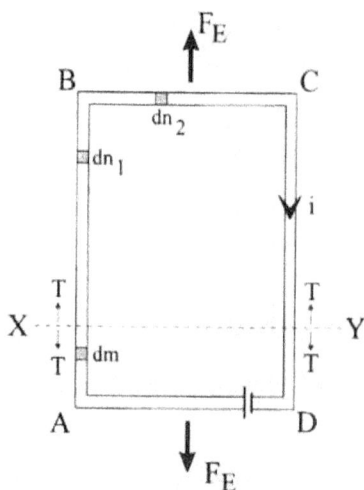

Figure 2.23 : Electrodynamic forces in a rectangular circuit

Since the circuit is known not to lift itself as a whole, we must have

$$F_E = 2 T \qquad (2.51)$$

Surprisingly, certain calculations with Grassmann's law lead to the same result. This is one more reason why it has been argued that the two force laws are equivalent. In other words, the Grassmann law, notwithstanding its very different appearance in Eq.1.56, appears at first inspection to be compatible with Newton's third law and Newtonian mechanics.

However closer inspection of this problem reveals that the Grassmann law cannot be used in the Newtonian framework if it is to predict the experimentally measured forces. Let ΔF_R stand for the relativistic (Grassmann) interaction of two current elements, while ΔF_N represents the Newtonian (Ampère) interaction of the same pair of elements, so that in general

$$\Delta F_R \neq \Delta F_N \qquad (2.52)$$

In relativistic electromagnetism, two current elements on the same straight line exert no force on each other. In all other cases the Grassmann force is perpendicular to the current element on which it acts. Hence no relativistic force exists between dm and dn_1 of figure 2.23, which could contribute to the tension T. One might expect, therefore, that the whole of the tension is the result of interactions between one element in AD and the other in BC, because here the forces on the two elements are equal and opposite. Calculation shows, however, that

$$\left| \sum_A^D \sum_B^C \Delta F_R \right| \ll F_E \qquad (2.53)$$

or the calculated force is much smaller than the measured force. The inescapable conclusion from this is that Grassmann's law, after all, is not compatible with Newtonian mechanics. In order to use Grassmann's law to calculate the measured force F_E, it must be used in accordance with a different system of mechanics, that of special relativity.

In relativistic mechanics and field theory it has to be assumed that the magnetic field at any current element is due to the circuit as a whole. When this is taken into account, the correct result

$$\left| \sum_{ABCD} \sum_B^C \Delta F_R \right| = F_E \qquad (2.54)$$

is obtained. It is as if a current element generates a magnetic field strength at another element, and then absolves itself from any responsibility with regard to the reaction force. In this way the current element $i.dn_1$ of figure 2.23 produces a magnetic field at $i.dn_2$, and this field then generates a lift force ΔF_R on $i.dn_2$, which contributes to F_E, without an elemental reaction force. This is the relativistic self-force mechanism, because dn_1 and dn_2 are integral parts of the solid body XBCY cut from the rectangular circuit and electrically reconnected across the surface X-Y by two thin liquid mercury films. Roper [2.24] actually performed measurements of F_E with liquid mercury films at X and Y.

From this it has to be concluded that, in order to arrive at the correct experimental result of Eq.2.51, calculations with the Grassmann formula must invoke the relativistic mechanics of self-forces. Hence the possible agreement of the two force laws on a particular prediction does not eliminate the need for the relativistic mechanics.

It does not follow that when using the two mechanics in their appropriate spheres of validity, they will always agree on the outcome of a specific experiment. For example, they predict different force distributions in the rectangular circuit of figure 2.23. This can be shown as follows.

Using the Newtonian electrodynamics with Ampère's force law, F_E has to be calculated according to the Slater-Frank rule

$$F_E = \left| \sum_{XADY} \sum_{XBCY} \Delta F_N \right| \qquad (2.55)$$

While performing this summation, step by step, it will be found that most of F_E exists in the form of longitudinal forces which have their seats near X and Y. In contrast to this, the Grassmann formula places all of the lifting force in the top leg BC of the circuit. In the Newtonian mechanics, the body XBCY is pushed upward from below, and in the relativistic mechanics, it is pulled up by a force on the uppermost part of the body. There exist experiments, such as the previously mentioned impulse pendulum, and others [1.12], which can distinguish between the two force distributions. The Ampèrian distribution has been confirmed and, therefore, the Grassmann distribution is incorrect.

The step-by-step summation of Eq.2.55 also reveals that the tension T in AB and CD does not disappear when a complete circuit is considered. Hence the earlier example of figure 2.2 was sufficient to prove that Ampère's law predicts tension in any straight wire section. The force law controversy has, therefore, been resolved by experiments. A long discussion of the differences between Newtonian and relativistic forces can also be found in our recent book [2.30].

In 1993, after the publication of the Robson-Sethian paper [2.26], Phipps announced and provided the formal proof of his shape-independence theorem [2.31]. When applied to the circuit of figure 2.24, this theorem states:

> "For the electrodynamic force laws (cf. Whittaker [1.6]) of Lorentz (Grassmann, Biot-Savart, Laplace, etc), Ampère-Weber, Gauss-Riemann, and all others differing only by additive exact differential quantities, the net longitudinal ponderomotive force component (if any) acting parallel to the length of a straight current-carrying test element, denoted armature A, exerted by currents flowing in a fixed external partial circuit C of given shape joining fixed endpoints E, E' (these points and C being nowhere coincident with A), is independent of the shape of C and depends only on the geometry of the gaps EA and E'A."

If we apply Phipps' theorem to the net longitudinal force on the armature A of figure 2.24, the theorem asserts that this force is independent of the shape of the remainder of the circuit, C. Calculations with Ampère's force law do indeed confirm this prediction.

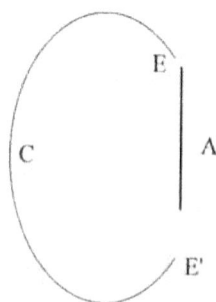

Figure 2.24 : Circuit used for demonstration of shape independence theorem

In order to test the theorem by experiment, arc gaps have to be provided on either end of the armature so that the current loop is continuous, and also so that the piece of conductor may accelerate longitudinally. One of the authors tested Phipps' theorem at the University of Oxford and found that, for equal length gaps, the armature would not accelerate longitudinally [2.32]. Unequal length gaps did, however, result in longitudinal acceleration in quantitative agreement with Ampère's force law when applied using the assumption that current element interactions between arc elements and armature elements can be ignored. The importance of these unequal gaps was ignored in the Robson-Sethian paper [2.26], and the fact that they only used equal gaps was the reason that they did not observe the longitudinal forces that they were looking for.

The neglect of electrodynamic forces on the solid armature due to the current in the arcs is justified by inertia considerations. The elements of the armature have two types of elemental interactions, which must be treated separately. The first is the interaction force between an arc element and an armature element, which will impart considerable acceleration to the almost free arc ion of minute mass (inertia), and very little acceleration to the armature element which, because of atomic bonding, carries the inertia burden of all of the armature atoms. This results in the extremely lopsided acceleration of the arc plasma and virtually no acceleration of the armature. However the second type of interaction forces are between one element in the armature and another in the rest of the metallic circuit, which is attached to the earth and is therefore much heavier than the armature. These forces are manifested almost totally as an acceleration of the armature. If the two arc gap lengths are unequal, then these forces, which are of the second type, on the armature do not necessarily sum to zero, thus leaving a net longitudinal force on it.

In the Oxford experiment, capacitors from 3.3 to 10 µF were charged to 33 kV and discharged through the circuit described in figure 2.25. Maximum current amplitudes ranged from 40 to 70 kA. They varied a little for the same capacitor charge because of statistical fluctuations of the pre-breakdown current. The 55 mm long, 6.35 mm diameter, copper armature was arranged vertically, and the initial bottom gap varied between 0 and 10.2 mm. The initial top gap variation was from 10.2 to 20.3 mm. The armature was allowed to move upward through a light friction grip, which completely prevented downward motion. As a result the upward armature displacement, h, could be measured. These values can be compared to the Ampère force prediction as shown in figure 2.26. The Ampère force in this case is calculated by ignoring the force interactions between current elements in the armature and those in the arc gap plasma. This is justified by the reasoning described earlier.

Finally two qualitative observations were recorded. The friction support of the armature was inverted so that the top gap could be made zero and downward motion was possible. After a medium current discharge, the armature was found to have crossed the gap, showing that in this position the armature was subject to a downward force, as predicted by Ampère's law. Perhaps in the most striking experiment, an aluminum armature of identical size was used and soldered to the bottom copper electrode. Aluminum was employed because it does not make a very strong bond with conventional solder. Nevertheless, the joint was strong enough so not to break by pulling lightly by hand. When the capacitor bank was discharged through the soldered joint, it broke and the armature crossed the gap. This proved conclusively that the longitudinal armature motion was not caused by arc pressure.

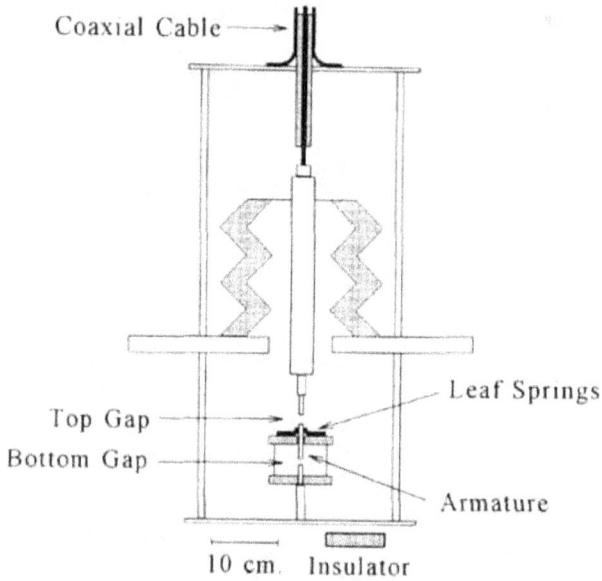

Figure 2.25 : Experimental arrangement for the Oxford longitudinal force demonstration

The comparison between the experimental data and the predicted Ampère force is shown in figure 2.26.

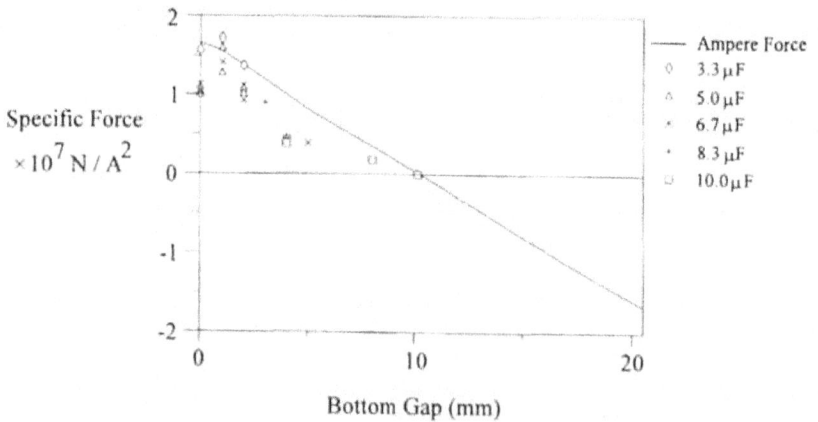

Figure 2.26 : Experimental data points and Ampère force prediction (solid curve) from Oxford experiment

The good agreement between the experimental data and the theoretical prediction, as well as the two qualitative tests leave no room for doubt that longitudinal Ampère forces do exist, and they can accelerate an armature which is located between unequal arc gaps. Purely transverse Lorentz forces could never explain this behaviour. This completes the description of longitudinal force demonstrations. Others have been performed and are recorded in the literature.

Chapter 2 References

2.1 A.M. Ampère, *Theorie mathematique des phenomenes electrodynamique uniquement deduite de l'experience*, Blanchard, Paris, 1958.

2.2 C. Blondel, *Ampère et la creation de l'electrodynamique*, Bibliotheque Nationale, Paris, 1982.

2.3 J. Nasilowski, "Phenomena connected with the disintegration of conductors overloaded by short-circuit currents", Prezeglad Electrotechniczny (Poland), Vol.37, No.10, p.397, 1961.

2.4 J. Nasilowski, "Unduloids and striated disintegration of wires". *Exploding wires*, (W.G. Chase, H.K. Moore, Editors), Vol.3, Plenum, New York, 1964, p.295.

2.5 J. Nasilowski, "Ampère tension in electric conductors". IEEE Transactions on Magnetics, Vol.MAG-20, p.2158, 1984.

2.6 P. Graneau, "First indication of Ampère tension in solid electric conductors", Physics Letters, Vol.97A, p.253, 1983.

2.7 P. Graneau, "Ampère tension in electric conductors". IEEE Transactions on Magnetics, Vol.MAG-20, p.444, 1984.

2.8 E.F. Northrup, "Some newly observed manifestations of forces in the interior of an electric conductor", Physical Review, Vol.24, p.474, 1907.

2.9 H. Aspden, "The exploding wire phenomenon". Physics Letters, Vol.107A, p.238, 1985.

2.10 H. Aspden, "The exploding wire phenomenon", Physics Letters A, Vol.120, p.80, 1987.

2.11 J.G. Ternan, "Stresses in rapidly heated wires", Physics Letters A, Vol.115, p.230, 1986.

2.12 P. Graneau, "Wire explosions", Physics Letters A, Vol.120, p.77, 1987.

2.13 C. Hering, "Electromagnetic forces: a search for more rational fundamentals; a proposed revision of the laws", Journal AIEE, Vol.42, p.139, 1923.

2.14 C. Hering, "Revision of some of the electromagnetic laws", Journal of the Franklin Institute, Vol.194, p.599, 1921.

2.15 P. Graneau, "Electromagnetic jet propulsion in the direction of current flow", Nature, Vol.295, p.311, 1982.

2.16 A.M. Hillas, "Electromagnetic jet propulsion: non-lorentzian forces on currents?", Nature, Vol.302, p.271, 1983.

2.17 P.T. Pappas, "The original Ampère force and Biot-Savart and Lorentz forces", Nuovo Cimento, Vol.76B, p.189, 1983.

2.18 P. Graneau, P.N. Graneau, "The electromagnetic impulse pendulum and momentum conservation", Nuovo Cimento, Vol.7D, p.31, 1986.

2.19 P. Graneau, P.N. Graneau, "Electrodynamic momentum measurements", Journal of Physics D: Applied Physics, Vol.21, p.1826, 1988.

2.20 W.H.K. Panofsky, M. Phillips, *Classical electricity and magnetism*, Addison-Wesley, Reading MA, 1962.

2.21 *American Institute of Physics Handbook*, McGraw-Hill, New York, 1957.

2.22 P. Graneau, "Longitudinal magnet forces?", Journal of Applied Physics, Vol.56, p.2598, 1984.

2.23 C. Christodoulides, (a) "Are longitudinal forces predicted in magnetostatics by the Ampère force law in its line-current-element form?", Physics Letters A, Vol.120, p.129, 1987; (b) "Equivalence of the Ampère and Biot-Savart force laws in magnetostatics", Journal of Physics A: Mathematical and General Physics, Vol.20, p.2037, 1987; (c) "Comparison of the Ampère and Biot-Savart magnetostatic force laws in their line-current-element forms", American Journal of Physics, Vol.56, p.357, 1988; (d) "On Ampère's magnetostatic force law and the nature of the forces it predicts", Physics Letters A, Vol.141, p.383, 1989.

2.24 J.W. Roper, "Experimental measurement of mechanical forces in electric circuits", Journal AIEE, Vol.46, p.913, 1927.

2.25 P. Graneau, "Electrodynamic seawater jet: An alternative to the propeller?", IEEE Transactions on Magnetics, Vol.25, p.3275, 1989.

2.26 A.E. Robson, J.D. Sethian, "Railgun recoil, Ampère tension and the laws of electrodynamics", American Journal of Physics, Vol.60, p.1111, 1992.

2.27 Editorial: "Sixty years of the American Journal of Physics - more memorable papers", American Journal of Physics, Vol.61, p.103, 1993.

2.28 P. Graneau, "The difference between Newtonian and relativistic forces", Foundations of Physics Letters, Vol.6, p.491, 1993.

2.29 J.C. Slater, N.H. Frank, *Mechanics*, McGraw-Hill, New York, 1947.

2.30 P. Graneau, N. Graneau, *Newton versus Einstein*, Carlton Press, New York, 1993.

2.31 T.E. Phipps, "Ampère tension and Newton's laws", Apeiron, No.17, Oct. 1993, p 1.

2.32 N. Graneau, T.E. Phipps, D. Roscoe, "Experimental confirmation of longitudinal electrodynamic forces", submitted for publication.

Theoretical Developments

Finite Current Element Analysis

More than anyone else, Maxwell was responsible for making the Ampère-Neumann theory obsolete, at least temporarily. At the same time he was also the finest scholar of the old electrodynamics and made major contributions to it in the form of the geometric-mean-distance method of inductance calculations. This chapter summarizes the various extensions of the Newtonian electrodynamics which have been made during and since the time of Maxwell. The review will not follow the historical sequence of events, but will deal with the most significant milestone first, for this makes it much easier to understand the rest.

This milestone was the evolution of digital computers. It made it possible to explore the predictions of the laws of Ampère and Neumann to a breadth and depth that remained hidden to the founders of Newtonian electromagnetism. It was not so much the amount of computing which became feasible, but rather the avoidance of integration difficulties which had reduced the theory to a qualitative tool. When working with current elements of finite size, the solution process became transparent and inspired confidence. Any mathematical technique which ignored the indivisibility of electric charge could be seen to be an artifact.

There can be no doubt that Ampère himself believed his electrodynamic forces acted directly on the conductor metal, rather than the subtle electric fluid contained in the metal. The Ampère forces were ponderomotive forces, just like the forces of gravitation. The matter that experienced the forces did not necessarily move and, therefore, Ampère's force law did not include a velocity. To explain electromagnetic induction, Neumann postulated a second and totally different set of forces. He called them electromotive forces (e.m.f.). The electromotive forces did not act on the metal, but on the electric fluid contained in the conductor. Since Neumann's law of induction, Eq. 1.39, followed directly from Ampère's force law, it seemed that the same material elements were involved in both laws.

Finite current element analysis may be used to calculate both ponderomotive and electromotive forces in metallic conductors and dense plasmas. It may be assumed that the element is simply a piece of metal containing electric charges. The charge configuration must permit current flow, caused by electromotive forces, to coexist with ponderomotive forces

91

Theoretical Developments

on the metal lattice. It has never been adequately explained how this is possible.

Many modern investigators still treat the electric current as being infinitely divisible, as Ampère did. This was quite acceptable in the 1820s before the atomicity of matter had become firmly rooted in our understanding of nature. If the current element is an element of matter, it is no longer reasonable to assume it to be infinitely small. In order for the metal to sustain the ponderomotive tension and compression forces which exist across atomic bonds, the Ampèrian current element must involve the atomic nucleus. The length of the atomic current element would then have to be limited to the spacing of atoms in the metal lattice. The shape of the current element could be taken to be that of the unit cell of the lattice. However the number of atoms in the lattice would be far too great to be handled by even the most powerful computers, and thus we were compelled to cluster atomic current elements to form macroscopic elements suitable for finite element analysis.

The infinitely small current elements proposed by Ampère could be handled with integral calculus. This turned out to be very convenient for calculating ponderomotive forces between two circuits which did not approach each other very closely. Difficulties arose when it became necessary to calculate the reaction forces between parts of the same circuit, as this involved the interaction of neighboring small elements. Due to its inverse square nature, the strongest contribution to Ampère tension actually arises from the force between adjacent elements where the integral diverges. This is the reason why so few calculations of longitudinal forces were published in the 170 years since Ampère formulated his law.

By not allowing element self-interactions, the integration singularities completely disappear when finite current elements are employed, even if the elements are as small as a unit cell of the lattice. This is why the availability of computers has contributed so much to the renewed interest in the Ampère-Neumann electrodynamics. These integration difficulties have helped to conceal the difference between the old and the new electrodynamics of metals.

In the analysis of two circuits made of wires of a diameter which is small compared to the distance between the circuits, almost any kind of resolution of the circuits into a reasonable number of elements will give similar answers for the mutual force or the mutual inductance. When the circuits nearly touch each other, the choice of current element shape and size greatly influences the computed results. Analytical solutions do not eliminate this difficulty. Take, for example, the mutual inductance of equal coaxial circles as computed by Maxwell with elliptic integrals and later tabulated by Grover [3.1]. The closest distance for which Grover lists a mutual inductance is four percent of the radius of the circle. At closer range his formula diverges so rapidly that interpolation becomes meaningless. For extremely close circles Grover recommends a logarithmic formula which could have easily arisen from finite element summations.

Element shape and size selection has the greatest effect on the calculation of selfinductances and reaction forces between parts of the same circuit. These calculations involve the interaction of elements that are *in contact* with each other. For this reason, most investigators have avoided the issue of element size. Let us look at a useful paper by Charles [3.2] for the computation of Lorentz forces on electric power conductors. This author uses what has become known as the "stick model". He calculates the forces between two or more straight conductor sections, or sticks, which do not touch each other. Each stick is treated as a single filament of current elements. Since the stick is straight, the elements within it exert no Lorentz forces on each other. Therefore, forces between neighboring elements do not enter the calculation. The stick model is unsuitable for Ampère's law because it predicts longitudinal force interactions within each stick.

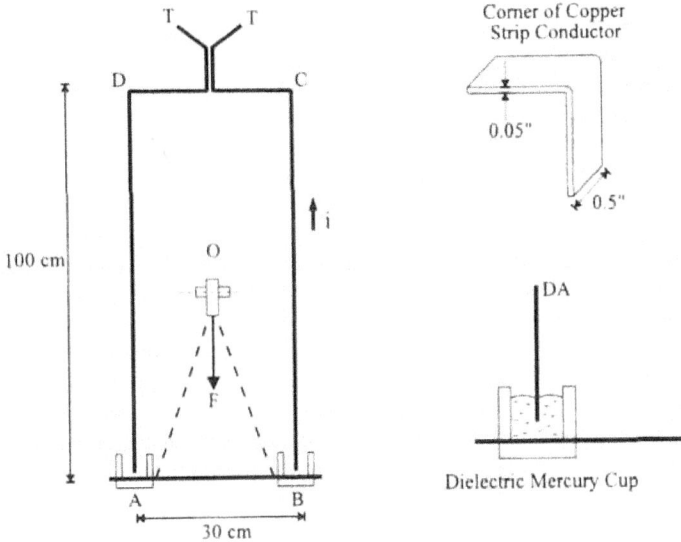

Figure 3.1 : Experimental arrangement for reaction force measurements on a rectangular circuit

Charles devoted a paragraph to the forces at bends and corners of a circuit. This is where two sticks have to meet. In this paragraph he says:

> "As x (distance from corner) tends to zero the current in the bend tapers off with a corresponding reduction in the mechanical forces in the vicinity of the corner. The problem is outside the scope of the paper, but an approximate solution may be obtained for a 90 degree bend by assuming that the force starts at the point x=0.779r, where r is the radius of the conductor."

Charles' small gap right where two sticks meet sharply modifies the computed forces acting on either stick. Cleveland [3.3] too was forced to resort to imaginary conductor gaps at the corners of his circuit to make his force calculations agree with experiment.

Consider a closed circuit made up of a wire of diameter d. For the purpose of calculating the reaction forces between two parts of the circuit, treat the wire as a single filament of elements. How long should the elements be? Neither Ampère nor Neumann provided guidelines to answer this question. With computers the authors have found that the finite current element method will yield reasonable results only if the element length is approximately equal to the wire diameter. In other words, the length to width ratio of the current element should be unity or close to it. If all elements are of this shape and equal size and completely fill the conductor volume, then the element length is also equal to the distance between adjacent elements. This statement implies that the position of an element is given by the geometric center of the element volume.

If a large conductor is thought of as consisting of a bundle of filaments, circular filament cross-sections would not fill the conductor volume. Thus for the subdivision of large conductors into a number of parallel filaments it is more appropriate to choose filaments of square cross-section. The requirement of unity length-to-width ratio then makes the current element a cube. Examples treated in the rest of the book will, therefore, be based on cubic elements. In a straight conductor the longitudinal and lateral spacing of adjacent elements may then be taken to be equal to the element length. To check how cubic current element computations compare with experiment, the apparatus of figure 3.1 has been used. It consists of a 100×30 cm rectangle made of 0.5 inch wide and 0.05 inch thick copper strip. The ABCD plane of the circuit was mounted vertically with the bottom side AB horizontal and hung at O from a beam balance. AB was electrically connected to the remaining sides of the circuit via non-metallic cups A and B filled with liquid mercury. In each cup the current had to cross a depth of about one millimeter of mercury. With the conductivity ratio of copper to liquid mercury being approximately fifty, the current will cross the gap almost along straight and parallel streamlines. Hence the corners at A and B, as well as those at C and D, were quite sharp, as indicated in figure 3.1. The current was led in and out of the circuit with closely packed parallel strips between C and D to terminals T-T.

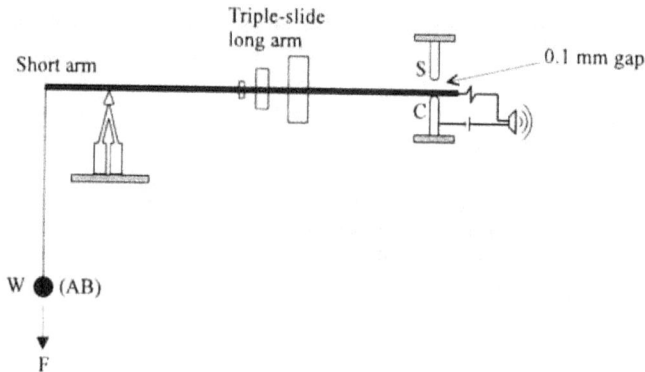

Figure 3.2 : Force balance

The transverse force on side AB acted vertically downward. It was measured with the beam balance with a sensitivity of 0.1 gram. Stabilized DC currents up to 500 A were passed around the circuit and gave rise to transverse forces on AB in the range from 100 - 250 mN. The force balance technique is further explained by figure 3.2. This is a simplified diagram of a commercial beam balance of unequal arms. The weight of side AB with mercury cups and support components will be denoted by W. This weight W, hanging on the short arm, was balanced with the aid of three graded weights on a calibrated triple-slide on the long arm. The balance condition was detected with a battery driven piezoelectric whistle. As shown in figure 3.2, the whistle circuit was completed through a mechanical contact between the long balance arm and the metallic support C.

Even when no current flows in the circuit ABCD, not all of W is dead weight. A small part is due to buoyancy forces in the mercury cups. Copper has a specific gravity of 8.94 while that of liquid mercury is 13.55. Therefore the ends of the vertical conductors dipping into the mercury push side AB down. Slight tipping of the balance during the experiment affects the immersion depths in the cups. If this depth is decreased by one millimeter, then the reduced buoyancy force leads to an underestimate of the electrodynamic force F by 1.5 mN. To keep this error small, the balance swing was limited to an angle just sufficient to break contact C. The positioning of the mechanical stop S, of figure 3.2, restricts the immersion level changes to a small fraction of one millimeter.

Another source of error is the thermal expansion of the 100 cm long vertical conductor portions of figure 3.1. This has been controlled to a certain extent by not allowing the conductor temperature to rise above $70°C$. Copper has a coefficient of linear thermal expansion of $16.42 \times 10^{-6} (°C)^{-1}$. The maximum thermal expansion of sides AB and BC therefore was 0.82 mm. The effect of this expansion is to increase the depth of immersion in the mercury cups. It counteracts the error due to the reduced buoyancy force caused by beam swinging.

The major measurement error arose from the fact that it seemed to be impossible to obtain a clean break of contact C for a force increment of 1 mN, the nominal sensitivity of the balance. The contact resistance at C was a function of contact pressure and this, in turn, determined the loudness of the whistle. The stabilizing effect of buoyancy changes, and possibly also elastic strain hindered clean breaking of contact C. For these various reasons the accuracy of the force measurement was unlikely to be better than ± 5 mN.

The measurements plotted on figure 3.3 were made by observing the following procedure.

(1)   With the current switched off, the sliding weights were adjusted until the contact C was just broken and whistling ceased, but for an occasional light bounce, say once per second, caused by floor-born vibrations. A check was made that in this condition the gap at S was clearly open.

(2)   The weights on the long beam were then adjusted to push the beam firmly down on C, producing a loud whistle tone. The value of the weight adjustment W was noted. This operation pre-loaded the balance negatively by W.

(3)   Next a preset DC current through the rectangular circuit was switched on for 30 seconds. If this did not interrupt the whistle sound, a higher current was pre-selected and the experiment was repeated. A note was made of the current I which just broke contact C according to the definition given under (1). The circuit was allowed to cool down between repeated current applications.

(4)   The balance was then pre-loaded to a different value of W and the

Theoretical Developments

corresponding value of I was determined by the method of (3). The weights W were taken to represent the vertical transverse electrodynamic force on circuit side AB.

The measurement points plotted on figure 3.3 show quite reasonable--but not perfect--proportionality of the force to the square of the current. The slope of the broken line, which is meant to represent the measurements, was found to be 9.85.

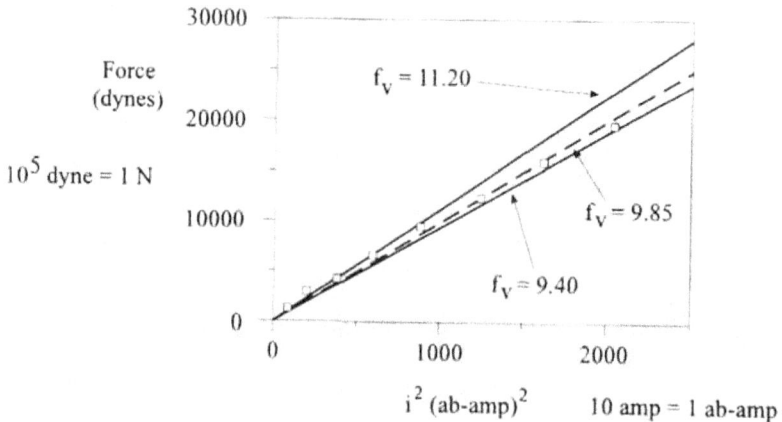

Figure 3.3 : Measured transverse force on AB

A finite current element analysis was carried out for two different element shapes shown in figure 3.4. In figure 3.4(a) the circuit was modeled as a single filament, carrying all of the current i, made up of 1 cm long elements. Each element then took the shape of a 0.05 inch thick, 1 cm × 0.5 inch plate. This made the ratio of the greatest width to element length approximately one. The other element representation is sketched in figure 3.4 (b) and (c). It involves the subdivision of the conductor strip into ten square-section filaments, each carrying i/10 ab-amps of current. Individual filaments are then further subdivided into 0.05 inch long cubic current elements. This resolved the circuit into 20,460 cubic elements as compared with the 260 plate-elements of the single filament model. The finer resolution of the conductor was expected to give the more accurate results.

The vertical downward force on AB is mainly due to the interaction of this side with AD and BC. With a single filament the problem is two-dimensional, while with ten filaments it becomes three-dimensional. We will immediately analyze the more complex three-dimensional case as this also explains how to handle the two-dimensional circuit. With respect to the 10-filament model, let the corner A be placed at the origin of a rectangular coordinate system, as indicated in figure 3.5. The y-coordinate labels the ten parallel filaments from y = 0 to y = 9. Once the interaction of an end-filament in AD and all filaments of AB has been calculated, the total interaction of the two sides of the rectangular circuit may be determined from a 10×10 filament interaction matrix.

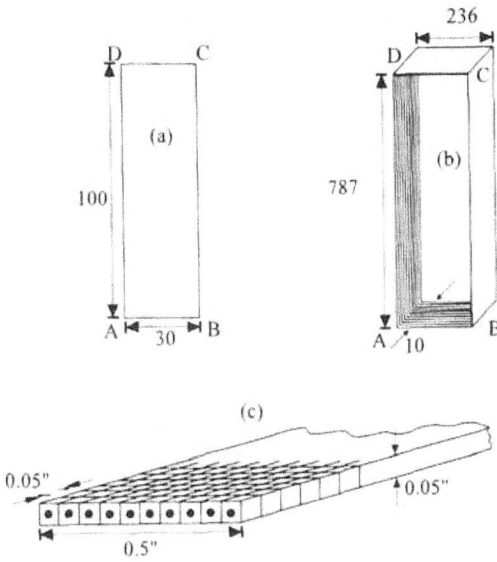

(a) single filament model
(b) 10 filament model
(c) strip of 10 filaments

Figure 3.4 : Current element representations

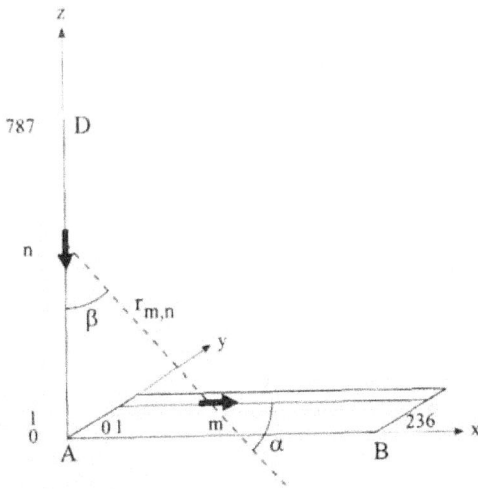

Figure 3.5 : Three-dimensional element interaction

Now consider the general element m (figure 3.5) in AB which has coordinates m $(x_m, y_m, 0)$ and the general element n in the first filament of AD which lies on the z-axis and has coordinates n $(0, 0, z_n)$. The distance between those two elements is given by

$$r_{m,n}^2 = (x_m - x_n)^2 + (y_m - y_n)^2 + (z_m - z_n)^2 = x_m^2 + y_m^2 + z_n^2 \qquad (3.1)$$

The required direction cosines of $r_{m,n}$ are

$$\cos \alpha = x_m / r_{m,n} \qquad (3.2)$$

$$\cos \beta = z_n / r_{m,n} \qquad (3.3)$$

It should be noted that for interactions between sides AB and AD the angle of inclination of the current elements is $\varepsilon=90°$ and therefore $\cos\varepsilon=0$. Then from Eq.1.24, the vertical downward component of the elemental Ampère force $\Delta F_{m,n}$ is

$$(\Delta F_{m,n})_v = - \left(\frac{i}{10}\right)^2 \frac{dm\ dn}{r_{m,n}^2} (-3\cos\alpha\ \cos\beta)\ \cos\beta$$

$$= 3 \left(\frac{i}{10}\right)^2 \frac{dm\ dn}{r_{m,n}^2} \cos\alpha\ \cos^2\beta \qquad (3.4)$$

which can be solved by substitution of Eqs.3.1-3.3. Equation 3.4 has to be summed for all combinations of m and n indicated by figure 3.5. This task can only be accomplished by a computer.

It is found that if all the current would flow in just one filament--say the filament coinciding with the coordinate axes of figure 3.5--the force between AD and AB would be significantly greater than in the case where the current is distributed over ten parallel filaments. This demonstrates that the total force is also a function of current density.

One of the advantages of writing the equations of the Ampère-Neumann electrodynamics in fundamental electromagnetic units (e.m.u.) is that the force is seen to be dimensionally equivalent to the square of a current. This is obvious from Ampère's force law Eq.1.24, and also from Eq.3.4. Therefore the force equations are independent of the unit of length. Length may be measured in meters, feet, or any other unit, without changing the equations. Finite current element analysis can be greatly simplified by making the chosen length of the current element also the unit of length. For this purpose we set

$$dm \ = \ dn \ = \ 1 \text{ unit of length} \tag{3.5}$$

Of course in any given problem all elements should be of the same length. With Eq.3.5 the distance between two current elements is expressed by a number of element lengths. The recurring ratio $(dm \ dn \ / \ r^2_{m,n})$ is thus a dimensionless number. Hence in fundamental electromagnetic units we may divide the force by the square of the current (i^2) and obtain another dimensionless number which will be referred to as the 'specific force' and noted as f.

The downward force on the yth filament in AB (see figure 3.5) due to the filament labelled 0 in AD is

$$F_{0,y} \ = \ 3 \left(\frac{i}{10} \right)^2 \ \sum_{m=1}^{236} \ \sum_{n=1}^{787} \ \frac{\cos \alpha \ \cos^2 \beta}{r^2_{m,n}} \tag{3.6}$$

This shows that AB = 30 cm has been resolved into 236 elements and AD = 100 cm into 787 elements. Furthermore, $r_{m,n}$ of Eq.3.6 must be equal to the number of elements that can be fitted into the distance between the general elements m and n.

The interaction of all ten filaments in AD with the ten filaments in AB can be compiled in a 10 × 10 symmetrical matrix of the form

$$
\begin{matrix}
F_{0,0} & F_{1,0} & F_{2,0} & \cdot & \cdot & \cdot & F_{9,0} \\
F_{0,1} & F_{1,1} & F_{2,1} & \cdot & & & F_{9,1} \\
F_{0,2} & F_{1,2} & F_{2,2} & \cdot & & & F_{9,2} \\
& & & & & & \\
& & & & & & \\
F_{0,9} & F_{1,9} & F_{2,9} & \cdot & \cdot & \cdot & F_{9,9}
\end{matrix}
$$

The total vertical force on AB due to AD is the sum of the elements of this matrix. Since it is symmetrical, and elements along any diagonal are all equal to each other, we find the force in question to be

$$
\begin{aligned}
F_v \ = \ & 10 \, F_{0,0} \ + \ 18 \, F_{0,1} \ + \ 16 \, F_{0,2} \ + \ 14 \, F_{0,3} \ + \ 12 \, F_{0,4} \\
& + \ 10 \, F_{0,5} \ + \ 8 \, F_{0,6} \ + \ 6 \, F_{0,7} \ + \ 4 \, F_{0,8} \ + \ 2 \, F_{0,9}
\end{aligned} \tag{3.7}
$$

It is convenient to solve Eq.3.6 for the specific force f_v such that

$$f_v \ = \ \frac{F_v}{i^2} \tag{3.8}$$

Equations 3.7 and 3.8 also give the force exerted by side BD of figure 3.4 on side AB in the downward direction because of the symmetry of the rectangular circuit.

Finally we have to compute the downward force exerted by CD on AB. For the wide separation of these two sides, the shape and size of the current elements chosen has little influence on the result. CD contributes less than two percent to the downward force on AB and force dilution due to the parallel filaments will be negligible. Hence a single-filament representation may be used for this side pair in which each side is resolved into 30, 1-cm-long, plate elements. Using this approximation, the specific downward force on AB due to CD came to 0.1724. The total downward force on AB due to the other three sides of the rectangle was thus found to be $f_v = F_v / i^2 = 9.40$, which has been plotted on the graph of figure 3.3.

Pinch forces in the mercury cups should exert a downward thrust on AB in each cup of $0.5i^2$. The pinch force analysis supporting this claim was carried out by Northrup [2.8]. With pinch thrust added to the finite element calculations, using cubic elements, the total specific downward force comes to 10.40 compared to the experimental value of 9.85. Taking experimental uncertainties into account, this example supports the use of macroscopic cubic current elements in finite element problems.

With the rectangular circuit of figure 3.4 being represented by just a single filament of 1 cm long elements, in the shape of thin plates, the calculated specific downward force came to 11.20. This has also been plotted on figure 3.3. It clearly is an overestimate of the specific force even when the pinch thrust is ignored. In view of the greatly reduced amount of computation, however, the single filament model is a worthwhile approximation.

The example of the rectangular circuit illustrates that the force on part of a circuit due to the remainder of the circuit is primarily determined by the interaction of neighboring elements. Remote parts of the circuit make only a small contribution. In many practical situations they may be ignored. The prominence of the local interactions is the result of the inverse square law in Ampère's formula, Eq.1.24.

In Chapter 2, the principles of finite current element analysis were applied to the single filament model of a square circuit to explain what is meant by Ampère tension. We will now delve deeper into this subject by studying a straight conductor which has been subdivided into a number of parallel filaments. First we ask the question: what will be the tension in two side-by-side filaments which share the current of one absolute ampere? To obtain an answer consider the two square-section filaments of figure 3.6. The two filaments have been subdivided into four portions a, b, c, and d. Each portion consists of $z/2$ cubic current elements with their vectors all pointing in the same direction. Let us now determine the specific tension T_t / i^2 across the midplane of the filament combination when each filament carries half the total current, or $i/2$. The tensile force due to the interaction of portions a and b can be derived directly from Eqs.2.4 or 2.16. An equal component will arise from the interaction of portions c and d. Let these two components be $T_{a,b}$ and $T_{c,d}$. Then from Eq.2.16

$$T_{a,b} = T_{c,d} = \frac{1}{4} (0.19 + \ln z) i^2 \qquad (3.9)$$

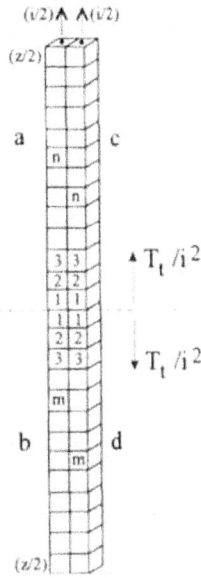

Figure 3.6 : Tension across the mid-plane of a filament pair

For calculation of $T_{a,d}$ and $T_{c,b}$ which, because of symmetry, are equal to each other, we find from figure 3.6 that

$$r_{m,n}^2 = (m + n - 1)^2 + 1 \tag{3.10}$$

$$\cos \varepsilon = 1 \tag{3.11}$$

$$\cos \alpha = \cos \beta = \frac{m + n - 1}{r_{m,n}} \tag{3.12}$$

Applying Ampère's force law, Eq.1.24, to portions a and d of the filament pair of figure 3.6 and resolving the elemental interaction force in the direction of the current, we obtain

$$T_{a,d} = T_{c,b} = \frac{1}{4} i^2 \sum_{m=1}^{z/2} \sum_{n=1}^{z/2} \frac{1}{r_{m,n}^2} (2 \cos \varepsilon - 3 \cos \alpha \cos \beta) \cos \alpha \tag{3.13}$$

Solving Eq.3.13 by computer for a range of z-values, and applying regression analysis to the results revealed the logarithmic relationship

$$T_{a,d} = T_{c,b} = \frac{1}{4}(-1.64 + \ln z)\, i^2 \tag{3.14}$$

Hence the total tension T_t across the midplane of the filament combination is

$$T_t = 2T_{a,b} + 2T_{a,d} = (-0.73 + \ln z)\, i^2 \tag{3.15}$$

which is smaller than the force given by Eq.2.16. This result demonstrates that the calculated Ampèrian tension will be reduced if the conductor is broken down into a larger number of filaments. This has been called force dilution.

The single filament representation of a straight conductor is the crudest model one can use. Finer subdivision of the conducting matter into smaller cubes should result in better approximations to the specific tension. How far must this process be driven to obtain meaningful results? To investigate this question let every element of figure 3.7(a) be subdivided into eight smaller cubes, as shown in (b) of the diagram. This simple subdivision multiplies the computational work by a factor of at least 64. Hence too fine a subdivision of the conductor can be costly in computing time. For figure 3.7(b) angles α and β are no longer zero for all relevant current element combinations and Eq.3.13 has to be used in addition to Eq.2.4.

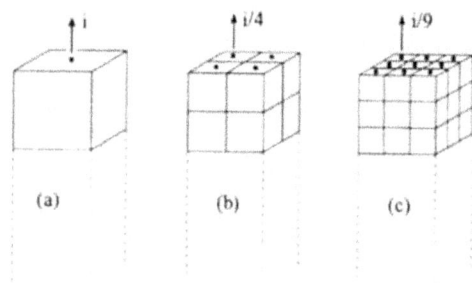

Figure 3.7 : Cubic element subdivision of a straight conductor

To obtain a quantitative indication of the magnitude of force dilution for successively thinner filaments, let us analyze a conductor of 2 m length and 1 cm² cross-section. The return circuit would make a contribution to the maximum tension in this conductor, but we will not compute this. Using finite element analysis, and the midplane tension was found to be

Number of filaments:	1	4	9	16	25	36	10^{14}
T_t/i^2	5.4893	4.7094	4.5479	4.4891	4.4612	4.4459	4.4122

This example indicates that the computed specific Ampère tension converges quite rapidly as the number of parallel filaments, n, is increased. In general, a plot of (T_t / i^2) vs. $1/n$ yields a straight line, whose y-intercept is the predicted tension with $n \sim 10^{14}$ atomic width filaments

Further examples of the use of finite current element analysis will be found in the remainder of the book.

Reaction Forces from the Selfinductance Gradient

The most frequently used formula for calculating the reaction forces between two parts of a current-carrying circuit, consisting of metallic conductors, is neither the Ampère nor the Lorentz force law, but

$$F_x = \pm \frac{1}{2} i^2 \frac{\partial L}{\partial x} \tag{3.16}$$

where ∂x is a virtual displacement between the two parts of the circuit in the direction x in which the reaction force F_x is required. As will now be shown, this is a formula of the Newtonian electrodynamics.

Let a complete circuit of selfinductance L carry a current i. Equation 3.16 is written in e.m.u. in which inductance has the dimension of length. The positive x-direction is the direction in which the circuit portions separate from each other by the virtual displacement. The $\pm$ sign signifies that Eq.3.16 is sometimes written with the negative sign. Whether the two reaction forces represent attraction or repulsion appears to be left to the judgment of the reader. In fundamental electromagnetic units the magnetic energy stored by the circuit is

$$E = \frac{1}{2} L i^2 \tag{3.17}$$

For two closed circuits m and n, Neumann's electrodynamic potential (stored magnetic energy) is given by Eq.1.25. It involves the mutual inductance, Eq.1.26, of the two circuits which may be viewed as a measure of the capacity of the two circuits to store mutual magnetic energy. Neumann's introduction of the concept of virtual work allowed him to relate the mutual force of attraction or repulsion of the two circuits by Eq.1.27. Hence this mutual force may be written

$$(F_{m,n})_x = - \frac{\partial}{\partial x} (i_m i_n M_{m,n}) \tag{3.18}$$

where $M_{m,n}$ is the mutual inductance of the circuits.

The capacity of complete circuits to store magnetic energy must derive from their electrodynamic interactions which are governed by Ampère's law, Eq.1.24, upon which

Neumann's theory was based. Consequently, each pair of current elements stores mutual magnetic energy in accordance with an elemental mutual inductance of

$$\Delta M_{m,n} = - \left(\frac{2 \cos \varepsilon - 3 \cos \alpha \cos \beta}{r_{m,n}} \right) dm \; dn \tag{3.19}$$

Equation 3.19 was first suggested by Peter Graneau [1.12] and has become responsible for some of the modern developments of the Newtonian electrodynamics. For unknown reasons Neumann did not assign a mutual inductance to a pair of conductor elements. Later it will be shown that Eq.3.19 predicts the existence of mutual torques as well as forces between the elements. If Neumann's virtual work concept is generally valid, as is widely believed, then these mutual torques must exist.

If Eq.3.19 is integrated around a closed circuit, the angle function may be replaced by $\cos \varepsilon$, because the remainder of this function is a total differential and integrates to zero. This was first proved by Neumann and subsequently by many other investigators. Hence Eq.3.19 is compatible with the negative form of Eq.1.26.

It follows logically that the selfinductance $L_{o,o}$ of an isolated closed circuit is composed of elemental contributions defined by Eq.3.19. In the case of selfinductances the angle function of Eq.3.19 must not be simplified to $\cos \varepsilon$, because the sum of the interactions of one element dm with all the other elements dn does not involve a closed loop integration. The selfinductance of the closed single filament loop, therefore, is given by

$$L_{o,o} = - \sum_m \sum_n \left(\frac{2 \cos \varepsilon - 3 \cos \alpha \cos \beta}{r_{m,n}} \right) dm \; dn \quad (m \neq n) \tag{3.20}$$

where m and n are sequentially numbered labels of the conductor elements of o. The angles of Eq.3.20 are the angles of Ampère's law as drawn in figure.1.3. Each pair of elements is allowed to interact twice because in each inductive interaction one element is the cause and the other feels the induced e.m.f.

Equation 3.20 is actually a single-filament formula of selfinductance which can be solved by finite element analysis. Let us evaluate it for a circle of radius R, as shown in figure 3.8.

The circle has been divided into z+1 elements of equal length which are labelled 0, 1, 2, ..., n, ..., z. Let dm be the 0-th and dn be the n-th element. With the symbols defined in figure 3.8 we set

$$dm = dn = R \; d\theta = \left(\frac{2 \pi}{z+1} \right) R \tag{3.21}$$

$$\theta_n = n \; d\theta \tag{3.22}$$

$$r_{m,n}^2 = 2 R^2 (1 - \cos \theta_n) \tag{3.23}$$

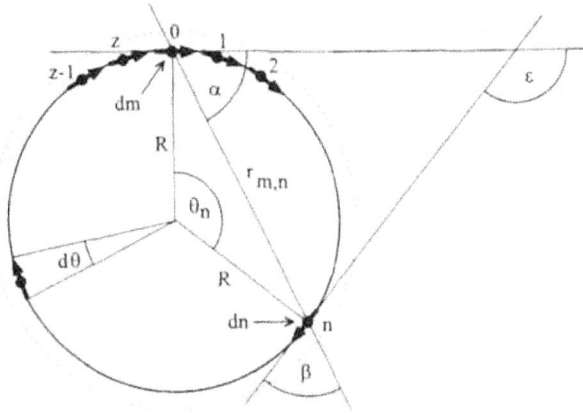

Figure 3.8 : Construction for the calculation of the selfinductance of a filamentary circle

The double sum of Eq.3.20 can then be reduced to a single sum

$$L_{o,o} = -(z+1) \ dm \ dn \ \sum_{n-1}^{z} \left(\frac{2 \cos \varepsilon - 3 \cos \alpha \cos \beta}{r_{m,n}} \right) \qquad (3.24)$$

because the symmetry of the circle ensures that the inductive interaction of dm with all other elements dn is the same wherever dm is situated. In Eq.3.24 each element pair effectively interacts twice. This is physically correct because of the one-way nature of inductive elemental interaction.

The relationship between the angles of figure 3.8 is as follows

$$\alpha = \beta, \ \varepsilon = \alpha + \beta = 2\alpha = 2\beta = \theta_n \qquad (3.25)$$

Finally, the symmetry of the circle permits us to sum only half-way around it and multiply the result by two. After making the necessary substitutions in Eq.3.24, the following formula was evaluated by computer for z ranging from 300 to 20,000

$$\frac{L_{o,o}}{R} = \frac{2\sqrt{2}\pi^2}{z+1} \sum_{n-1}^{z/2} \left(\frac{3 - \cos(n \ d\theta)}{\sqrt{1 - \cos(n \ d\theta)}} \right) \qquad (3.26)$$

In fundamental electromagnetic units this is a dimensionless shape constant. The computed results are listed in table 3.1. Regression analysis revealed that the figures closely obey the logarithmic formula

$$\frac{L_{o,o}}{R} = 13.835 + 12.599 \ \ln(z+1) \tag{3.27}$$

z	$L_{o,o}/R$ from: Eq.3.26	$L_{o,o}/R$ from: Eq.3.27
300	85.665	85.697
400	89.310	89.321
500	92.133	92.133
600	94.438	94.430
1000	100.887	100.866
2000	109.624	109.599
3000	114.729	114.707
5000	121.157	121.143
10,000	129.875	129.876
15,000	134.972	134.985
20,000	138.589	138.609
10^6		187.897
10^{12}		361.958

Table 3.1 $L_{o,o}/R$ of a circular filament

This leads to the conclusion that z must not approach infinity, or the filament must not be infinitely thin, as the Ampère-Neumann electrodynamics would then become meaningless. In any case, since the reaction forces are exerted on matter which is not infinitely divisible, it is only reasonable that the current element size should have a finite lower limit. In the absence of any better information, we have assumed that the lower limit is the atomic spacing, or approximately 10 Angstrom. For single filament circles ranging in radius from 1 mm to 1000 m, this implies 10^6 to 10^{12} elements per circle. The $L_{o,o}/R$ ratios of these two values were calculated and are listed in table 3.1. The figures are not unreasonably large. The fact that $L_{o,o}/R$ increases with z also complies with the known fact that the selfinductance of a circle of constant radius increases when the wire diameter is reduced.

Now that we have a method of calculating the selfinductance of a closed filament -- and this need not be a circle -- we return to the evaluation of the reaction force between two parts of an arbitrarily shaped circuit o by the virtual work equation, Eq.3.16. To do this it is necessary to determine the selfinductance gradient along some specified direction x. Consider the particular example illustrated in figure 3.9 The current elements are labelled from 1 to g along circuit portion ABC and from g+1 to g+h along the remaining portion C'DA'. The sums in the selfinductance equation, Eq.3.20, have to proceed from m=1 to m=g+h and from n=1 to n=g+h. The various summations used to calculate the selfinductance of the circuit of figure 3.9 can be arranged as

$$\text{If} \quad S = \left(\frac{2 \cos \varepsilon \; - \; 3 \cos \alpha \cos \beta}{r_{m,n}} \right), \quad \text{then}$$

$$L_{0,0} = - \left[\sum_{m=1}^{g} \sum_{n=g+1}^{g+h} S \, dm \, dn \; + \; \sum_{m=g+1}^{g+h} \sum_{n=1}^{g} S \, dm \, dn \; + \; \sum_{m=1}^{g} \sum_{n=1}^{g} S \, dm \, dn \right.$$

$$\left. + \; \sum_{m=g+1}^{g+h} \sum_{n=g+1}^{g+h} S \, dm \, dn \right] \qquad (3.28)$$

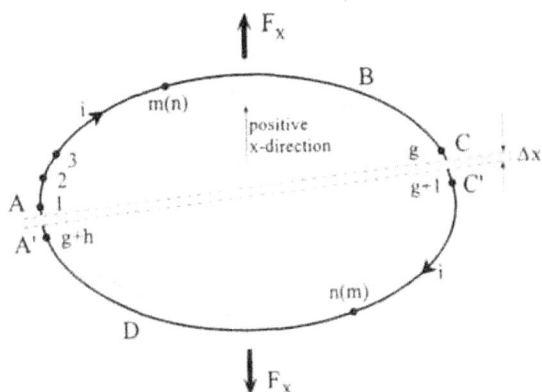

Figure 3.9 : Virtual displacement between parts of a filamentary circuit

How much will the selfinductance change when the two circuit portions of figure 3.9 are separated by the virtual displacement Δx ? Let this change be

$$(\Delta L)_{\Delta x} = L_2 - L_1 \qquad (3.29)$$

The third and fourth terms of Eq.3.28 are not affected by the virtual displacement. They remain constant and, therefore, drop out of the difference equation, Eq.3.29. It should also be noted that the first term of Eq.3.28 is equal to the second term and the two terms may be combined.

Potential energy and mutual inductance are scalar quantities. The magnetic energy storage capability, which is the inductance, does not change when the position of two conductor elements is interchanged. Hence the first and second terms of Eq.3.28 are not of opposite sign. The selfinductance change for the virtual displacement is then the difference in the limiting values of the first term of Eq.3.28, that is

$$(\Delta L)_{\Delta x} = -2 \sum_{m-1}^{g} \sum_{n-g-1}^{g-h} \left(\frac{2\cos\varepsilon - 3\cos\alpha\cos\beta}{r_{m,n}} \right) dm \; dn \; \Big|_{x-0}^{x-\Delta x} \qquad (3.30)$$

In Eq.3.30, m is confined to ABC and n to CDA' of figure 3.9. Let the distance $r_{m,n}$ between the general elements m and n after the virtual displacement be r_2, while before it was r_1.

Now consider the (2 cosε - 3 cosα cosβ) angle function. Figure 3.10 indicates that the parallel shift of element $i_m$dm along the direction x leaves the angle ε unchanged, however angles α_1 and β_1 become α_2 and β_2.

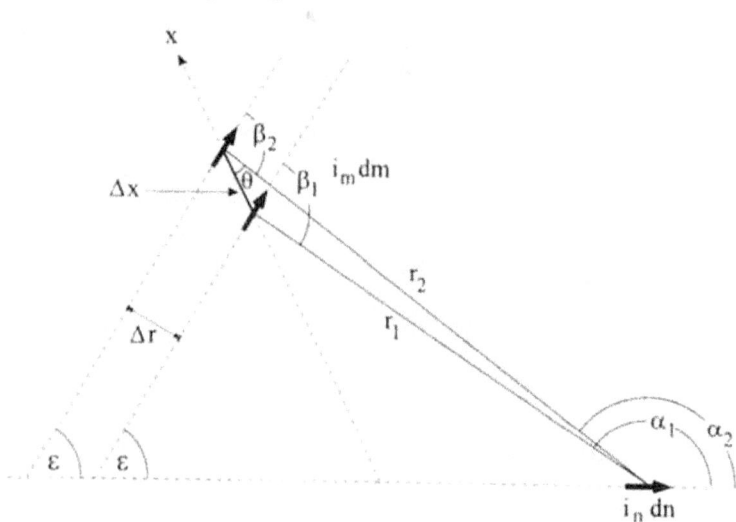

Figure 3.10 Virtual displacement between two general current elements

Next we note that

$$\alpha + \beta = \varepsilon + 180° = \text{constant}$$

Now examine the change in the product $(\cos\alpha \, \cos\beta)$. Because of the constant sum of the two angles we may also express them as

$$\alpha_2 = \alpha_1 - \Delta\alpha, \qquad \beta_2 = \beta_1 + \Delta\alpha$$

Then

$$\cos\alpha_1 \cos\beta_1 - \cos\alpha_2 \cos\beta_2 = \tfrac{1}{2}[\cos(\alpha_1 - \beta_1) - \cos(\alpha_2 - \beta_2)]$$

$$= \tfrac{1}{2}[\cos(\alpha_1 - \beta_1) - \cos(\alpha_1 - \beta_1)\cos(2\Delta\alpha) - \sin(\alpha_1 - \beta_1)\sin(2\Delta\alpha)]$$

Therefore, in the limit in which $\Delta x \rightarrow \partial x$ and $\Delta\alpha \rightarrow 0$, the second term becomes equal to the first and the last term vanishes. Hence the change in the angle function $2\cos\varepsilon - 3\cos\alpha\cos\beta$ tends to zero as the virtual displacement tends to zero. The change is, in fact, a first-order infinitely small quantity. This is sufficient for it to be negligible compared with the other quantities of Eq.3.30, that is dm, dn and $r_{m,n}$, which are of finite size. It would, of course, be insufficient in an integration over infinitely small current elements.

To a first approximation Eq.3.30 may, therefore, be written

$$(\Delta L)_{\Delta x} = 2 \sum_{m=1}^{g} \sum_{n=g+1}^{g+h} (2\cos\varepsilon - 3\cos\alpha\cos\beta)\left(\frac{r_2 - r_1}{r_1 \, r_2}\right) dm \ dn \qquad (3.31)$$

As will be seen from figure 3.10, for a small virtual displacement ∂x,

$$r_1 \, r_2 = r_{m,n}^2 \qquad (3.32)$$

$$r_2 - r_1 = \partial r = \partial x \, \cos\theta \qquad (3.33)$$

where θ is the angle of inclination of the x-direction to the $r_{m,n}$ direction. Substituting Eqs.3.32 and 3.33 into Eq.3.31 and dividing by ∂x gives

$$\frac{\partial L}{\partial x} = \lim_{\Delta x \to 0} \frac{\Delta L}{\Delta x} = 2 \sum_{m=1}^{g} \sum_{n=g+1}^{g+h} \left(\frac{2\cos\varepsilon - 3\cos\alpha\cos\beta}{r_{m,n}^2}\right) \cos\theta \ dm \ dn \qquad (3.34)$$

Now consider the reaction force between two circuit portions in the x-direction according to Ampère's law, Eq.1.24. It comes to

$$(F_{m,n})_x = -i^2 \sum_{m-1}^{g} \sum_{n-g-1}^{g-h} \left(\frac{2\cos\varepsilon - 3\cos\alpha\cos\beta}{r_{m,n}^2} \right) \cos\theta \ dm \ dn \qquad (3.35)$$

By comparing this last equation with Eq.3.34, it is seen that the reaction force may be written

$$(F_{m,n})_x = -\frac{1}{2}i^2 \frac{\partial L}{\partial x} \qquad (3.36)$$

By comparison of Eq.3.36 with Eq.3.16, it can be seen that the presently used force equation based on the selfinductance gradient comes from Neumann's virtual work concept. Equation 3.36 removes the uncertainty about the sign of Eq.3.16. To be consistent, the negative sign has to be chosen for all virtual work formulae.

This concludes the demonstration that both the Ampère reaction force calculations and the selfinductance gradient are based on Newtonian electrodynamics.

Relationship between Self and Mutual Inductance

Selfinduction is generally interpreted as a special case of mutual induction with primary and secondary circuit merged into one conductor. This view is supported by the fact that both quantities have the same dimension which in e.m.u. is length. It is the natural dimension of mutual inductance because this parameter appears to depend solely on the length and disposition of lines in space. In contrast to this, conventional selfinductance formulae apply to three-dimensional conductors. Measurements also show selfinductance to be a function of such non-geometrical quantities as resistivity, material homogeneity, and energizing frequency.

The concepts of self and mutual inductance arose in the explanation of Faraday's discovery of electromagnetic induction. Ampère did not take part in this effort. It was Neumann who derived the fundamental mutual inductance formula Eq.1.26, but he barely referred to selfinductance. In the closing remarks of his book [1.13] there is mention of the "extracurrent". By this he meant the induced current arising from switching currents on and off in an isolated circuit. The term selfinductance was probably coined by Kelvin [3.4] in order to define the total inductively stored energy in two circuits. Maxwell [1.8] greatly illuminated the meaning of the two inductance parameters, but his efforts in this direction also reveal a certain amount of confusion.

On the one hand Maxwell believed the magnetic manifestations of the electric current to be kinetic effects, and on the other he developed the geometric-mean-distance (GMD) method of computing inductances which is firmly rooted in Neumann's non-kinetic potential theory. To underpin his kinetic current model, Maxwell [1.8] built a mechanical machine in which two flywheels, representing primary and secondary currents, are connected through a differential gear to a third rotating member which makes "mutual inductance" possible. This machine is being preserved in the Cavendish Laboratory in Cambridge, England. As a result of this many scientists now treat selfinductance as a kind of electrical inertia and mutual inductance as the gearing together of selfinductances. Maxwell's GMD method of inductance calculation is embedded in the following treatment of selfinductance.

Consider a wire loop as shown in figure 3.11, part of which may take the form of a coil or solenoid. The loop current i is assumed to be driven by an externally generated e.m.f., E. At this stage no restriction need be placed upon the shape or size of the loop, nor the conductor cross-section and homogeneity, or on the rate of change of the applied e.m.f., so long as no charge accumulation takes place anywhere along the conductor.

If R is the resistance of the loop and L its selfinductance, then the loop current i is determined by the well-known equation

$$ iR = E - \frac{d}{dt}(Li) \qquad (3.37) $$

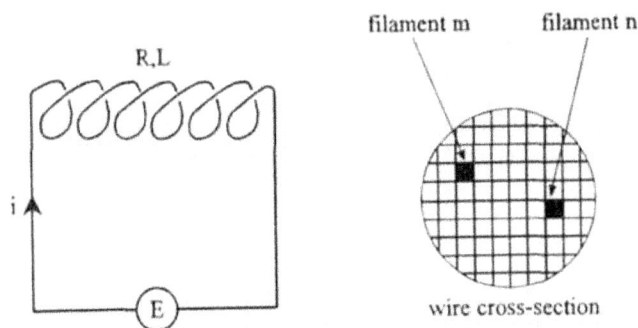

Figure 3.11 : Wire loop in series with an external e.m.f.

Let us now look at two general filaments m and n of the wire cross-section. All filaments must be thin tubes of flow filled with conducting matter. In figure 3.11 the filaments have square cross-sections, but any other shape could have been chosen provided this left no empty space between the filaments. According to the classical definition of a tube of flow, its cross-sectional area may vary along its length, but each cross-section must carry the same current. The current i_m flowing in filament m may be calculated from

$$ i_m R_m = E - \sum_n \frac{d}{dt}(M_{m,n} i_n) \qquad (3.38) $$

where R_m is the resistance of the m-th filament and i_n is the current in the n-th filament, while $M_{m,n}$ is the mutual inductance between the two general filaments. The summation in Eq.3.38 covers all possible positions of n in the wire cross-section, including that position in which n coincides with m. This coincidence defines the selfinductance of an individual filament as

$$L_m = M_{m,m} \qquad (3.39)$$

Bearing in mind that

$$\frac{1}{R} = \sum_m \frac{1}{R_m} \qquad (3.40)$$

$$i = \sum_m i_m = \sum_n i_n \qquad (3.41)$$

Equation 3.38 can be solved for i_m. Summing this solution for all filament currents in accordance with Eq.3.41, and remembering that resistance is not time dependent, gives

$$i = E \sum_m \left(\frac{1}{R_m} \right) - \frac{d}{dt} \left(\sum_m \frac{1}{R_m} \sum_n (M_{m,n} i_n) \right) \qquad (3.42)$$

Substitution of Eq.3.40 into Eq.3.42 and multiplication by R results in

$$iR = E - \frac{d}{dt} \left(\sum_m \frac{R}{R_m} \sum_n (M_{m,n} i_n) \right) \qquad (3.43)$$

A comparison of Eq.3.43 with Eq.3.37 defines the selfinductance of the wire loop as

$$L = \sum_m \frac{R}{R_m} \sum_n \left(M_{m,n} \frac{i_m}{i} \right) \qquad (3.44)$$

This is the most general and exact expression of selfinductance in terms of the filament self and mutual inductances of which the closed circuit is composed. The resistance ratio in Eq.3.44 explains why selfinductance is a function of material properties.

The expression for L takes on a more simple form when dealing with a wire of constant cross-section and uniform conductivity and which is also very long compared to its cross-sectional dimension. All filaments are then substantially of the same length and they may all be made to have the same cross-section. The resistance and current ratios of Eq.3.44 then depend merely on the total number of filaments. If this number is g, we have

$$\frac{R}{R_m} = \frac{i_m}{i} = \frac{1}{g} \qquad (3.45)$$

Equation 3.45 holds only if the rate of change of the applied voltage is sufficiently small for the current distribution over the conductor section to remain substantially uniform. This is why selfinductance depends on the energizing frequency which largely determines the current distribution.

Using Eq.3.45 the expression for the loop selfinductance reduces to

$$L = \frac{1}{g^2} \sum_m \sum_n M_{m,n} \qquad (3.46)$$

In this form L is seen to be the mean of all possible mutual inductance permutations of the g filaments, including a total of g combinations in which the positions of m and n coincide.

Maxwell [1.8] recognized that the mutual inductance of a pair of parallel straight lines is largely a function of the logarithm of the distance of separation d. For the purpose of mutual inductance calculations he further assumed that each conductor filament of finite thickness could be represented by a line coinciding with the filament axis. Then the average value of all of the mutual inductances making up a long straight conductor is determined by the average value of ln d. Since there are g filaments involved we have to deal with g(g-1)/2 different mutual inductances. A geometric mean distance (GMD) d' for all the filament combinations may now be defined as

$$\ln d' = \frac{2}{g(g-1)} \sum_{g(g-1)2} \ln d \qquad (3.47)$$

Once the GMD of the conductor cross-section has been found, it becomes possible to equate the selfinductance of this conductor, in accordance with Eq.3.46, to the mutual inductance of a single pair of lines separated by d'. Maxwell showed how to compute the GMD of a variety of conductor cross-sections. Furthermore, his GMD technique may equally well be applied to the mutual inductance of a pair of conductors of finite section, with each of them being subdivided into filaments. This latter computation requires the GMD of one conductor cross-section from the other, while the selfinductance calculation with Eq.3.46 depends on the geometric mean distance of an area from itself.

Maxwell [1.8] determined the GMDs for the most useful conductor configurations. The GMD of a circular area of radius r comes to 0.7788r, and that of a square area of side a, to 0.44705a. The GMD of a two-wire line is equal to the axial spacing of the conductors. The conductors have to be round wires, rods or tubes, but they do not have to be of the same diameter. In the case of a coaxial cable in which the outer tube has inner and outer radii of r_1 and r_0 respectively, the GMD, d', is given by

$$\ln\ d\ ' = \frac{r_o^2 \ln r_o - r_i^2 \ln r_i}{r_o^2 - r_i^2} - \frac{1}{2} \tag{3.48}$$

The shape and size of the inner conductor does not influence d'. Today it is possible to calculate the GMD for any conductor shape by computer assisted finite element analysis using Eq.3.47.

Maxwell's GMD method continues to be indispensable for practical inductance calculations. It is now often forgotten that, strictly speaking, it is valid only for very long straight conductors. Even in this restricted domain it involves Sommerfeld's approximation [3.5]. Sommerfeld was the first to solve Neumann's mutual inductance formula, Eq.1.26, for a pair of parallel straight filaments of finite length ℓ and spacing d. His result was

$$M_{m,n} = 2\left[\ell\ \ln\left(\frac{\ell + \sqrt{\ell^2 + d^2}}{d}\right) - \sqrt{\ell^2 + d^2} + d\right] \tag{3.49}$$

When ℓ is very much greater than d, Eq.3.49 simplifies to Sommerfeld's approximation

$$\frac{M_{m,n}}{2\ell} = -1 + \ln(2\ell) - \ln d \tag{3.50}$$

Only in this last equation is the mutual inductance per unit length proportional to the logarithm of the reciprocal of d, as assumed by Maxwell.

Maxwell's GMD method also ignores the selfinductance of individual filaments. Similarly, applying Neumann's formula, Eq.1.26, to two coinciding filaments gives an infinite result which must be meaningless. This is of fundamental importance because it shows that filaments used in inductance calculations must be of finite size. It makes eminent physical sense when they are elements of matter. There is no reason to believe that the elements of inductance calculations are not the very same elements which sustain the ponderomotive forces of Ampère's law. In other words, we are again dealing with metal atoms. Therefore, in the Newtonian electrodynamics, inductance is a property of matter and subject to the non-continuum nature of matter.

If we do not wish to follow Maxwell, who ignored the selfinductance of individual filaments, we have to assign a mutual inductance to a pair of elements, as was done in Eq.3.19. This becomes important when, for economy in computer time, the conductor is subdivided into a small number of filaments and the shape of the loop is such that the GMD method gives too coarse an approximation. The finite element determination of the selfinductance of a single filament has already been explained in conjunction with the circle of figure 3.8 and will be further treated in the following section. With this addition to the theory it has become clear that electromagnetic induction is an interaction of individual conductor elements. Whether or not the two elements which interact belong to the same or different circuits is not important. This disposes of the notion that selfinduction is somehow different from mutual induction. The idea of the existence of a mutual inductance of two elements, however, cannot be reconciled with the flux linkage concept of inductance in field theory.

Inductance of Single-Filament Circuits

A filamentary circuit is taken to be one in which the conductor cross-section is so small that forces and inductances may be calculated to adequate accuracy without further subdivision of the conducting area. Knowledge of the filament selfinductance is required for calculating self and mutual inductances of relatively thick conductors. The selfinductance defined by Eq.3.46 depends on the sum of the elements of a g × g mutual inductance matrix which contains the filament selfinductances along the principal diagonal. In fact the diagonal components will be greater than any other inductances in the matrix. This makes filament selfinductances particularly important when g is small.

The selfinductance of a closed filamentary circuit is given by Eq.3.20. This formula was applied to a filament circle and gave its selfinductance per unit radius by the logarithmic law Eq.3.27, involving the number of elements of which the filament was composed. The individual conductor (filament) element should be approximately as long as it is wide. This rule determines the number of elements z contained in the filamentary circuit.

Self and mutual inductances, based on the elemental mutual inductance, Eq.3.19, may be assigned to separate portions of circuits without loss of physical meaning. Particularly useful are straight-line filaments of finite length, which can later be combined to make up complete circuits.

Let a straight filament section of length ℓ be subdivided into z equal-length conductor elements. Then all inductive interactions within this section can be listed on a symmetrical square matrix of order z × z, with zeros along the principal diagonal. The zeros arise from the fact that in Newtonian theories an individual element does not interact with itself. The selfinductance of the filament section is the sum of all of the matrix elements. According to Eq.3.19, two similarly directed conductor elements, dm and dn, lying on the same straight line, separated by $r_{m,n}$, have a mutual inductance of

$$\Delta M_{m,n} = \frac{dm\ dn}{r_{m,n}} \tag{3.51}$$

The interaction of all neighbor-element combinations spaced dn apart will be found in the second diagonals, adjacent to the principal diagonal. The neighbor pairs all have the same mutual inductance of dm dn/dn=dm, and there exist 2(z-1) of them. The third diagonals contain the mutual inductances of pairs spaced two element-lengths apart. Their magnitudes are dm/2, and there are 2(z-2) of them. Continuing in this way from diagonal to diagonal, the selfinductance of the straight filament may be written

$$L(\text{straight}) = dm \ dn \ 2 \left[\frac{z-1}{dn} + \frac{z-2}{2 \ dn} + \frac{z-3}{3 \ dn} + \dots \right] = dm \ 2 \sum_{x=1}^{z-1} \frac{z-x}{x} \qquad (3.52)$$

Now the length ℓ of the filament is

$$\ell = z \ dm = z \ dn \qquad (3.53)$$

so that

$$\frac{L(\text{straight})}{\ell} = \frac{2}{z} \sum_{x=1}^{z-1} \frac{z-x}{x} \qquad (3.54)$$

This summing operation has been performed for eight values of z from 100 to 40,000. The results are reproduced in table 3.2 and figure 3.12. As expected, they obey a logarithmic relationship. Regression analysis has shown this to be

$$\frac{L(\text{straight})}{\ell} = -0.84 + 2.00 \ \ln z \qquad (3.55)$$

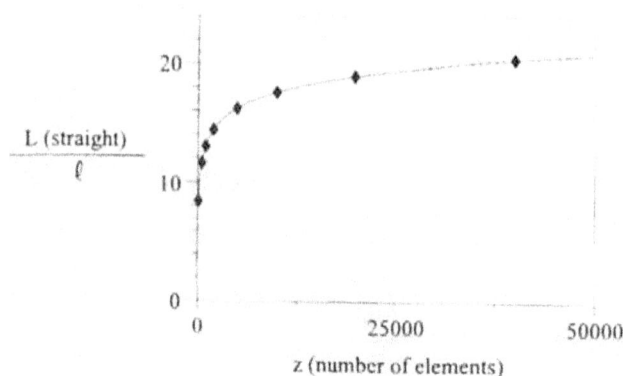

Figure 3.12 Selfinductance per unit length of a straight filament; curve from Eq 3.55; points from Eq 3.54

Extrapolations of Eq.3.55 to $z=10^6$ and 10^{12} are also noted in table 3.2. For a fundamental element length of ten Angstrom, the two large z-values correspond to conductor filaments of 1 mm and 1 km length, respectively, which comprises most practical applications.

If we treat a wire as a single filament then, for a fixed length, an increase in z implies a reduction in wire diameter. Figure 3.12 may therefore be used to determine the relationship between the selfinductance of a straight wire and its diameter. Alternatively, if the wire diameter is held constant, the increase in z represents a proportional increase in wire length. The selfinductance per unit length of a given wire is seen to increase logarithmically with length and never tends to a limit, however long the wire may be.

z	L(straight)/ z from Eq. (3.54)	from Eq. 3.55
100	8.375	8.370
500	11.586	11.589
1000	12.971	12.976
2000	14.357	14.362
5000	16.189	16.194
10,000	17.575	17.581
20,000	18.961	18.967
40,000	20.348	20.353
10^6		26.791
10^{12}		54.422

Table 3.2 Selfinductance per unit length of a straight filament

Knowledge of the selfinductance of rings, circular turns, and helically wound circular solenoids is of great importance in many areas of electromagnetic engineering. Maxwell addressed this subject with the GMD technique which has endured to this day. It is by no means perfect, nor easy, but it was the best one could do until computers became available. As Maxwell himself pointed out, the GMD method is rigorous only in the case of straight conductors, but it will furnish good approximate results when the cross-sectional dimension of the conductor, or the winding of a number of turns, is small compared to the diameter of the ring or the solenoid. Just how small the dimensional ratio must be for the approximation to hold, Maxwell did not specify.

To examine this last point we compare the mutual inductance of a pair of straight and parallel filaments with that of a pair of coaxial circular filaments, the two pairs having the same length and spacing, as indicated in figure 3.13.

The mutual inductance of the straight filament pair is given by Sommerfeld's equation, Eq.3.49 and approximation Eq.3.50. It was Maxwell who solved Neumann's mutual inductance formula, Eq.1.26, for two coaxial circles of radius r_1 and r_2 and a separation d between the planes of the circles. The solution takes the form

$$M_c = 4\pi \sqrt{r_1 r_2} \left(\left(\frac{2}{k} - k \right) K - \left(\frac{2}{k} \right) E \right) \tag{3.56}$$

where K and E are complete elliptic integrals of the first and second kind of modulus k. This modulus is

$$k = \frac{2\sqrt{r_1 r_2}}{\sqrt{(r_1 + r_2)^2 + d^2}} \tag{3.57}$$

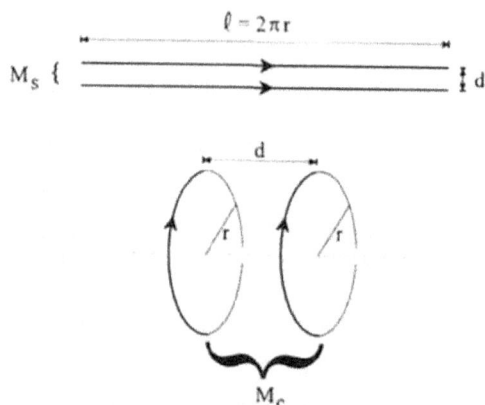

Figure 3.13 : Comparison of the mutual inductance of a pair of straight filaments with a pair of circular filaments of the same length and spacing

Tables of elliptic integrals are readily available but interpolation between tabulated values is frequently inadequate. Grover [3.1] reports that no less than 100 series expansions and other formulas have been published to overcome this interpolation difficulty. For two circles of radius r, separated by a small distance d, Grover cites a logarithmic relationship corresponding to

$$\frac{M_c}{2\pi r} = \ln\left(4\left(\frac{r}{d}\right)^2 + 1\right) - 1.2275 \tag{3.58}$$

As an alternative, figure 3.14 has been constructed to compute the mutual inductance of two equal coaxial filament circles by the finite element method. Each circle is divided into z equal arc-elements numbered from 0 to z-1. The element dm is held fixed in position m=0, while dn is taken around the circle n. Any individual position of dn is described by the angle ε, which is related to n through

$$\varepsilon = \frac{2\pi n}{z} \tag{3.59}$$

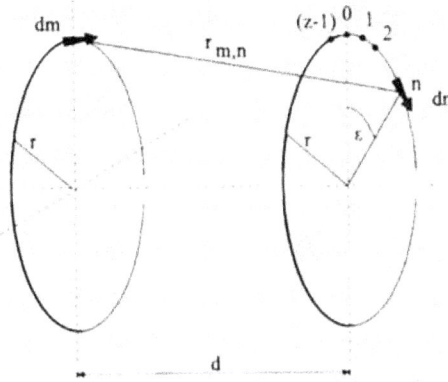

Figure 3.14 Construction for the solution of Eq. 3.63

The conductor element length is best written in terms of the radius r

$$dm = dn = \frac{2 \pi r}{z} \qquad (3.60)$$

The distance between dm and dn can then be shown to be

$$r_{m,n} = r \sqrt{\left(\frac{d}{r}\right)^2 + 2\left(1 - \cos\left(\frac{2 \pi n}{z}\right)\right)} \qquad (3.61)$$

Therefore, with Eq.3.19, the mutual inductance between dm and the whole of circle n is

$$dm \sum_{n=0}^{z-1} \frac{\cos \varepsilon}{r_{m,n}} dn \qquad (3.62)$$

This quantity is the same regardless of where dm is situated on the circle m. Hence the mutual inductance between the two circles is simply z times Eq.3.62. Substituting appropriately for the various parameters, this result is expressed by

$$M_c = \frac{4 \pi^2 r}{z} \sum_{n=0}^{z-1} \frac{\cos\left(\frac{2 \pi n}{z}\right)}{\sqrt{\left(\frac{d}{r}\right)^2 + 2\left(1 - \cos\left(\frac{2 \pi n}{z}\right)\right)}} \qquad (3.63)$$

Equation 3.63 tends to a finite limit as z is increased. Fortunately it reaches this limit quite quickly unless the circles are extremely close together. Table 3.3 shows the convergence for two r/d ratios.

$r/d = 1$		$r/d = 20$	
z	M_c (cm)	z	M_c (cm)
2	10.9116	20	53.4477
4	5.4558	50	40.3413
6	4.9999	100	38.7651
8	4.9482	200	38.6721
10	4.9417	300	38.6717
20	4.9408	400	38.6717
30	4.9408	500	38.6717
40	4.9408		
50	4.9408		

Table 3.3 : Convergence of the mutual inductance of coaxial filament circles ($r = 1$ cm)

The data plotted on figure 3.15 compares the mutual inductance per unit length of a pair of parallel straight filaments (curve 1, from Eq.3.50) with the mutual inductance per unit length of a pair of filament circles of the same length as the straight filaments (curve 2). Curve 2 demonstrates excellent agreement between the finite element formula, Eq.3.63 (points) with Grover's logarithmic formula, Eq.3.58 and Maxwell's solution, Eq.3.56 (line).

A regression analysis of the computer data obtained with Eq.3.63 for the interval $10 \le (r/d) \le 100$ produced the following logarithmic formula for the finite element analysis of the circles

$$\frac{M_c}{2 \pi r} = 0.18 + 1.99 \ln (r/d) \qquad (3.64)$$

When substituting numerical values in the appropriate range into Eq.3.58 and Eq.3.64 it will be found that these two logarithmic formulas are almost identical.

Returning to the question of the value of r/d at which the GMD method applied to filament circles becomes unreliable, first of all it should be noted in figure 3.15 that up to $(r/d) = 100$, the mutual inductance per unit length of two circles differs appreciably from that of straight filaments. The percentage gap between the two quantities decreases as r/d becomes larger. However for $d=10^{-7}$ cm (i.e. atomic spacing) and $2 \pi r = 100$ cm it is only down to eight percent. In view of this finding it is unrealistic to state, as Maxwell did, that the two quantities can be equated to each other. It must be remembered that Maxwell did not have the benefit of knowing Sommerfeld's formula, Eq.3.49.

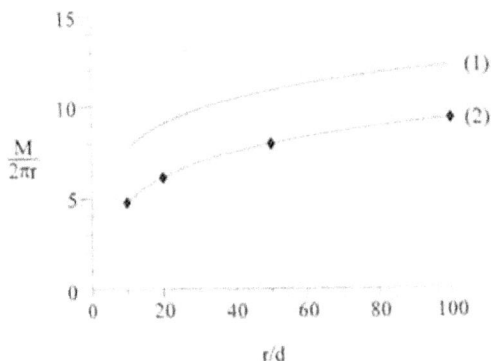

(1) parallel, straight filaments, Eq.3.50

(2) coaxial filament circles; curve (Grover's solution) Eq.3.58, and (Maxwell's solution) Eq.3.56 ; points (finite element analysis) Eq.3.63

Figure 3.15 Mutual inductance per unit length for the two geometries shown in figure 3.13

For very large r/d ratios both the elliptic integral and the finite element solutions agree with Eq.3.64, and thus the GMD method may be applied to conductor circles. In this case the selfinductance of a circular ring made, for example, of a round wire of radius 'a' may be computed with Eq.3.64 so long as the GMD of the wire cross-section (=0.7788a) is substituted for d.

How large must r/GMD be for this approximation to hold? It certainly would not hold when r=GMD=1 cm, for Eq.3.64 would then give a mutual inductance of 1.13 cm while the finite element method (table 3.3) indicates a value of 4.94 cm. Analysis of the finite element results reveals that the predicted mutual inductance only follows a logarithmic relationship such as Eq.3.64 as long as r/d ≥ 10. It is therefore suggested that the GMD method applied to conductor circles cannot be relied upon unless the r/GMD ratio is at least ten.

The finite element method of computing the selfinductance of an isolated conductor filament may be applied to any circuit geometry. Computing times strongly depend on circuit symmetry. To show some of the complexity that may arise, the filament square of figure 3.16 will be taken as the final example.

The sides of the square have been labelled (1) to (4) and each is divided into z equal-length conductor elements, so that the length of one side is given by

$$a = z \, dm = z \, dn \qquad (3.65)$$

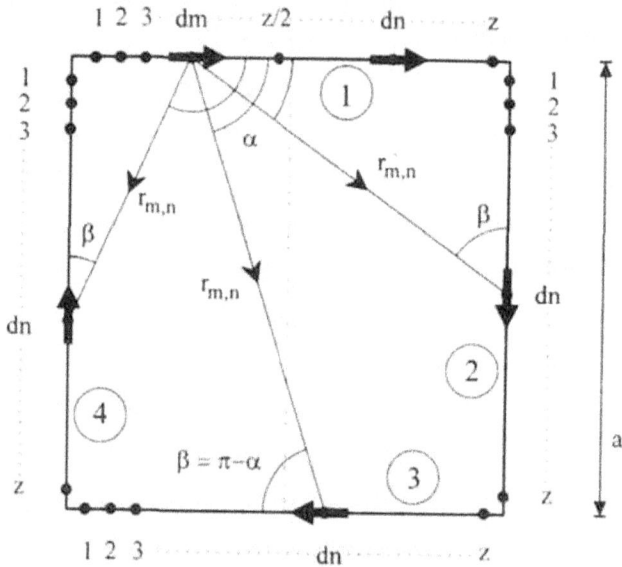

Figure 3.16 Construction for filament square

Element dm successively occupies positions 1 to $z/2$ along side (1). The inductance contributions from the remaining positions of dm may be deduced from the symmetry of the square. In fact the elemental inductance matrix will be of the order of $(4z \times 4z)$. In this matrix the mutual inductance of each element pair is listed twice. Now if dm moves from 1 to $z/2$ on side (1) while dn travels around the whole square, $((z/2) \times 4z = 2z^2)$ positions in the inductance matrix will be filled. A similar area of the matrix will be covered if dm moves from $(z/2)+1$ to z, and the sum of the respective elements will be identical to that obtained in the first operation. By repeating the process for dm on the remaining three sides, the whole of the inductance matrix may be filled. Symmetry permits us to write

$$
L(\text{square}) = 8 \sum_{m=1}^{z/2} dm \sum_{n=1}^{z} \left[\left(F(\alpha,\beta,\varepsilon)\frac{dn}{r_{m,n}} \right)_{S1} + \left(F(\alpha,\beta,\varepsilon)\frac{dn}{r_{m,n}} \right)_{S2} \right.
$$

$$
\left. + \left(F(\alpha,\beta,\varepsilon)\frac{dn}{r_{m,n}} \right)_{S3} + \left(F(\alpha,\beta,\varepsilon)\frac{dn}{r_{m,n}} \right)_{S4} \right] \quad (3.66)
$$

where S1 represents the interaction of dm with dn elements in side (1) and so on. With the construction of figure 3.16, and using Eq.3.20, the individual terms of the n-sum of Eq.3.66 are equal to

$$\left(F(\alpha,\beta,\varepsilon) \frac{dn}{r_{m,n}} \right)_{S1} = \frac{1}{|m - n|} \qquad (m \neq n) \qquad (3.67)$$

$$\left(F(\alpha,\beta,\varepsilon) \frac{dn}{r_{m,n}} \right)_{S2} = \frac{3(z+\frac{1}{2}-m)(n-\frac{1}{2})}{[(z+\frac{1}{2}-m)^2 + (n-\frac{1}{2})^2]^{1.5}} \qquad (3.68)$$

$$\left(F(\alpha,\beta,\varepsilon) \frac{dn}{r_{m,n}} \right)_{S3} = \frac{2}{[(m-n)^2+z^2]^{0.5}} - \frac{3(m-n)^2}{[(m-n)^2+z^2]^{1.5}} \qquad (3.69)$$

$$\left(F(\alpha,\beta,\varepsilon) \frac{dn}{r_{m,n}} \right)_{S4} = \frac{3(m-\frac{1}{2})(n-\frac{1}{2})}{[(m-\frac{1}{2})^2 + (n-\frac{1}{2})^2]^{1.5}} \qquad (3.70)$$

Finally, since dm = a/z, the dimensionless selfinductance per unit periphery of the square becomes

$$\frac{L(square)}{8a} = \frac{1}{z} \sum_{m=1}^{z \cdot 2} \sum_{n=1}^{z} \left[\frac{1}{|m-n|} + \frac{3(z+\frac{1}{2}-m)(n-\frac{1}{2})}{[(z+\frac{1}{2}-m)^2+(n-\frac{1}{2})^2]^{1.5}} \right.$$

$$\left. + \frac{2}{[(m-n)^2+z^2]^{0.5}} - \frac{3(m-n)^2}{[(m-n)^2+z^2]^{1.5}} + \frac{3(m-\frac{1}{2})(n-\frac{1}{2})}{[(m-\frac{1}{2})^2+(n-\frac{1}{2})^2]^{1.5}} \right] \qquad (3.71)$$

When solving Eq.3.71, the first term in the summation is ignored when m=n to avoid element self-interaction. Computer solutions for five values of z from 20 to 100 are listed in table 3.4.

As in previous examples, the filament selfinductance per unit length again obeyed a logarithmic relationship. For the square this was found to be

$$\frac{L(\text{square})}{8a} = 2.235 + 0.974 \ln z \tag{3.72}$$

The extent of the agreement between Eq.3.71 and Eq.3.72 is also indicated in table 3.4.

z	L(square) / 8a	
	by Eq.3.71	by Eq.3.72
20	5.1572	5.153
40	5.8247	5.828
60	6.2216	6.223
80	6.5050	6.503
100	6.7255	6.720
10^6		15.691
10^{12}		29.148

Table 3.4 : Selfinductance per unit length of a filament square

Inductance of Straight Conductors and Cables

Circuit self and mutual inductances are useful only if the current distribution over the conducting cross-section is reasonably uniform. Approximate uniform current distribution can always be assumed where the use of single filaments is justified. This has been the major reason for developing single-filament formulae.

One cause of non-uniform current distribution in homogeneous conductors is differences in filament lengths that arise, for example, in solenoids and all curved conductor segments. The second cause of current concentration in certain parts of the conductor is skin effect phenomena due to the flow of alternating or pulse currents. As the inductance of an AC or pulse current circuit determines the induced back-e.m.f., generated by the time-varying current, inductance formulae become of considerable importance. Few are available that can adequately handle non-uniform current distributions.

There is also interest in the inductance of large section conductors which carry DC or low frequency AC currents. Power conductors for the transmission and distribution of electricity fall into this category. The inductances may be required for the calculation of reactances, stored magnetic energy, and electrodynamic forces.

Although a straight conductor of finite length does not form a closed circuit, in the Newtonian electrodynamics it is permissible and expedient to associate it with force, inductance, and stored magnetic energy. The selfinductance of a homogeneous finite-length straight conductor is given by Eq.3.46. This rather simple equation implies that each place in the square matrix of filament mutual inductances contains a value, and therefore each mutual inductance appears twice in the double summation, once on either side of the principal diagonal, thus asserting that the selfinductance of a conductor is the average of the filament self and mutual inductances.

It is convenient to work with inductances per unit length. An appropriate formula for this purpose is Eq.3.55. To comply with the rules of finite element analysis, the number of elements, z, in each filament of length ℓ, must be

$$z = \frac{\ell}{dm} = \frac{\ell}{dn} \qquad (3.73)$$

Sommerfeld's solution, Eq.3.49 should be used to calculate the filament mutual inductances per unit length, which takes the form

$$\frac{M_{m,n}}{\ell} = 2\left[\ln\left(\frac{\ell}{d} + \sqrt{1 + \left(\frac{\ell}{d}\right)^2}\right) - \sqrt{1 + \left(\frac{d}{\ell}\right)^2} + \frac{d}{\ell}\right] \qquad (3.74)$$

The amount of computation involved in determining the selfinductance of the straight conductor depends on the total number of filaments g into which the conductor has been resolved. Surprisingly, relatively few filaments give quite accurate results. This will be illustrated with the 10×10 cm square-section conductor of figure 3.17 which is 1 km long and has been subdivided into g square-section filaments.

Computer results for this conductor are listed in table 3.5 for g varying between 100 and 10,000.

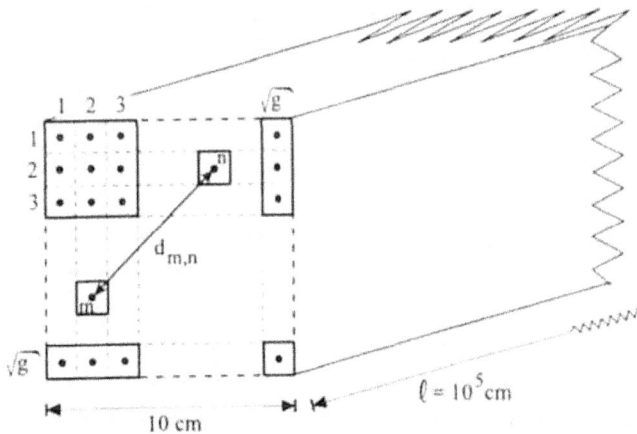

Figure 3.17 : Square-section conductor resolved into g filaments

Theoretical Developments

g	L/ℓ
100	19.3992
400	19.4128
900	19.4153
2500	19.4165
10000	19.4171

Table 3.5 : Selfinductance per unit length of a 1 km long straight conductor of 10 × 10 cm square cross-section

Multiplication of the number of filaments by 100 adjusts the L/ℓ ratio by less than 0.1 percent. Hence, at least in this example, 100 filaments must be giving a fairly accurate result. It should be observed that while the selfinductance per unit length of a 1 km long and straight filament ($z=10^{12}$) is 54.422, the selfinductance of the whole conductor is significantly smaller. Even though the length-to-width ratio of this conductor is 10,000, it cannot be adequately represented by a single filament, thus resolution into too few filaments will overestimate the selfinductance. The data of table 3.5 does not obey a logarithmic law.

The GMD of a square area of side 'a' is 0.44705a. Equating this to d in Eq.3.74 makes the selfinductance per unit length of the 10 × 10 cm, 1 km long conductor equal to 19.4172. This is remarkably close to the value listed in table 3.5 for g=10,000. It is probably the value which would be reached if the filament subdivision were driven to the atomic limit. The GMD method represents by far the easiest way of calculating the selfinductance of a straight conductor when the GMD of the conductor section is known. For any other conductor shape the finite element approach illustrated by figure 3.17 may be used with confidence.

The equivalence of the GMD and finite element techniques may also be shown as follows. When d/ℓ <<1, the non-logarithmic part of Eq.3.74 is very nearly equal to -1 and constant, that is independent of the actual length of the conductor. Sommerfeld's solution, Eq.3.74 then reduces to the approximation

$$\frac{M_{m,n}}{\ell} \approx 2(-1 + \ln 2 + \ln \ell - \ln d)$$ (3.75)

Since the selfinductance is equal to the average mutual inductance of the filament combinations we may write

$$\frac{L}{\ell} = \left(\frac{M_{m,n}}{\ell}\right)_{av} = -2 + 2\ln 2 + 2\ln \ell - 2(\ln d)_{av}$$ (3.76)

where $(\ln d)_{av}$ is the average value of ln d, and may be equated to

$$(\ln d)_{av} = \ln(GMD)$$ (3.77)

which proves the compatibility of the two methods of computation.

The Sommerfeld approximation, Eq.3.75 is not valid when d/ℓ cannot be ignored compared to 1. In such cases the average mutual inductance has to be determined by the much longer finite element process. To obtain an idea over what range of d/ℓ the approximation, Eq.3.75 may not be adequate, the logarithmic part and the remainder of Eq.3.74 have been computed with d/ℓ ranging from 1 to 10^{-5}. This revealed that Sommerfeld's approximation becomes unreliable when $d/\ell > 10^{-2}$. As a rule of thumb, therefore, the GMD method should not be used unless the conductor is more than 100 times as long as it is wide.

A straight conductor arrangement of great importance is the parallel go-and-return circuit. When the distance between the conductors is less than one percent of their length, the end connections closing the circuit have little effect on the total inductance and may be ignored. While studying the selfinductance of a single conductor, the direction of current flow was of no consequence, so long as it was the same in each conductor element. This is not the case in a go-and-return circuit. It has to be remembered that Sommerfeld's solution, Eqs.3.49 or 3.74, of Neumann's mutual inductance formula, Eq.1.26, assumes that dm and dn point in the same direction. If one of them is reversed, cos ε of every element combination changes from +1 to -1 and Eqs.3.49 and 3.74 change sign.

Consider the go-and-return circuit of figure 3.18. Each of the two conductors is of the same cross-section and made of the same homogeneous material. In this symmetrical situation it is convenient to subdivide each conductor into the same number of equal-area filaments, g.

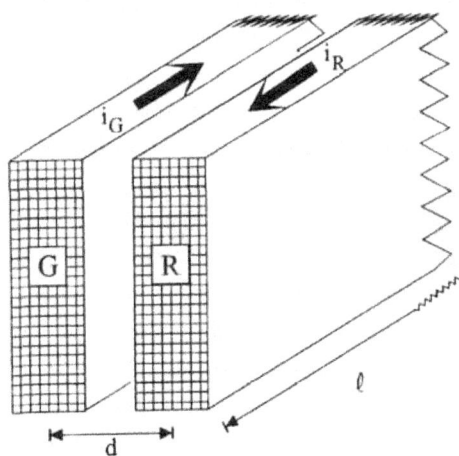

Figure 3.18 : Multi-filament model of a symmetrical Go-and-Return circuit

Figure 3.19 is the complete $2g \times 2g$ mutual inductance matrix with each filament mutual inductance occurring twice.

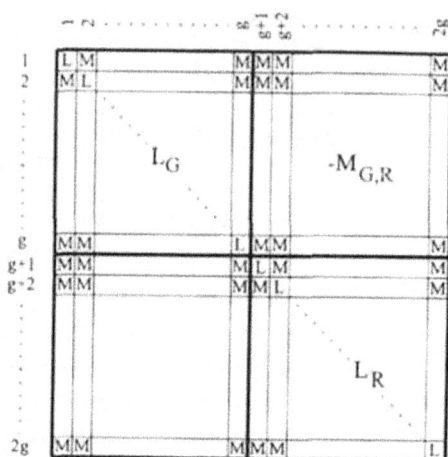

Figure 3.19 : Mutual inductance matrix for the symmetrical Go-and-Return circuit shown in figure 3.18

The form of the elemental mutual inductance equation, Eq.3.19, may be used as an analogy in order to express the energy stored by any pair of filaments m and n as

$$\Delta P_{m,n} = M_{m,n} i_f^2 \qquad (3.78)$$

where the filament current is defined by

$$i_f = \frac{i_G}{g} = \frac{i_R}{g} \qquad (3.79)$$

If the subscript G-R stands for the complete go-and-return circuit, the total stored energy may be written

$$P_{G-R} = \frac{1}{2} L_{G-R} i_G^2 = \frac{1}{2} i_f^2 \sum_m \sum_n M_{m,n} \qquad (3.80)$$

The double summation in Eq.3.80 must comprise all of the elements in the matrix of figure 3.19, however it is divided by two because the stored energy is represented by Eq.3.78. This division by 2 ignores the fact that filament selfinductances are only in the matrix once. However, this approximation becomes more accurate as the number of filaments is increased. According to Eqs.3.79 and 3.80 the selfinductance of the go-and-return circuit is

$$L_{G-R} = \frac{1}{g^2} \sum_m \sum_n M_{m,n} \qquad (3.81)$$

The matrix may be partitioned into four quarters, and with these partial matrixes Eq.3.81 may be split into

$$
L_{G\,R} = \frac{1}{g^2} \left[\sum_{m=1}^{g} \sum_{n=1}^{g} M_{m,n} - \sum_{m=1}^{g} \sum_{n=g+1}^{2g} M_{m,n} - \sum_{m=g+1}^{2g} \sum_{n=1}^{g} M_{m,n} \right.
$$

$$
\left. + \sum_{m=g+1}^{2g} \sum_{n=g+1}^{2g} M_{m,n} \right] \tag{3.82}
$$

It will be recognized that the first term of Eq.3.82 is the selfinductance L_G of the conductor G and the last term is L_R. The middle terms are equal to each other and represent the mutual inductance between G and R. The values of $M_{m,n}$ in the middle terms are negative because the currents in the two filaments are in opposite directions, which is why these terms are subtracted in the overall inductance calculation. Hence Eq.3.82 is equivalent to

$$
L_{G\,R} = L_G + L_R - 2 M_{G,R} \tag{3.83}
$$

An equation like Eq.3.83 applies to any go-and-return circuit, regardless of symmetry. If each conductor is resolved into a different number of filaments, then the mutual inductance matrix corresponding to that of figure 3.19 has to be partitioned into two squares of different size and two rectangles. The same method may be applied to the three conductors of a three-phase power transmission circuit [3.7]. Equation 3.83 is particularly useful when the circuit is long compared with the distance between the conductors so that the GMD method may be used for determining L_G, L_R, and $M_{G,R}$.

By way of an example, take two round conductors of radius r and separated from each other by the axial spacing d. The GMD of the circular area is 0.7788r and the GMD between two circular areas is equal to the axial spacing d. Then with Sommerfeld's approximation, Eq.3.75 and Eq.3.83 we obtain

$$
\frac{L_{G\,R}}{\ell} = 4 \left[-1 + \ln 2 + \ln \ell - \ln (0.7788\ r) \right] - 4 \left[-1 + \ln 2 + \ln \ell - \ln d \right]
$$

$$
= 4 \ln \left(\frac{d}{0.7788\ r} \right) \tag{3.84}
$$

The selfinductance formula normally quoted for parallel wire lines agrees with Eq.3.84 except that sometimes the logarithm is approximated to $\ln(d/r)$. In the development of Eq.3.84, the $\ln \ell$ terms cancel, demonstrating that the selfinductance per unit length of a go-and-return circuit does not depend on circuit length. In contrast to this, the selfinductance per unit length of an isolated straight conductor is length-dependent, as can be seen from Eq.3.76.

Maxwell [1.8] devised a useful theorem relating to the GMD between two areas A and X, when X can be split up in sub-areas X_1, X_2 ... X_n, in such a way that the GMD's between A and the sub-areas are known to be x_1, x_2, ... x_n. The theorem states that the GMD between A and X is

$$\ln (GMD) = \frac{X_1 \ln x_1 + X_2 \ln x_2 + \quad + X_n \ln x_n}{X} \tag{3.85}$$

where

$$X = X_1 + X_2 + \quad + X_n \tag{3.86}$$

Transient and Alternating Currents in Linear Conductors

The term linear conductor is meant to imply that the current streamlines, and therefore the conductor filaments, are straight and parallel. The mutual inductance between filaments is then given by Sommerfeld's solution, Eq.3.49, and the selfinductance may be calculated with Eq.3.55. The current distribution over the cross-section of linear conductors is of interest in pulse and AC power technology.

Maxwell [1.8] was the first to address the question of the distribution of time-varying currents over the cross-section of linear conductors. He spoke of cylindrical conductors and the difference in current intensity in the various cylindrical strata. This was the beginning of the skin effect theory which was further developed by Rayleigh [3.8] and others. Skin effect equations derived from field theory have been solved only for circular section conductors. In the case of non-circular conductors, the analytical approach fails because of the lack of knowledge of the appropriate boundary conditions. This problem with skin effect phenomena -- and its impact on Joule heating, magnetic energy storage, and force distribution -- was first overcome by Silvester [3.9] with a computer assisted finite element method.

We start by considering the irregularly shaped conductor of figure 3.20. This has been subdivided into a total of g equal-area square filaments. The two general filaments, m and n, are separated by the distance $d_{m,n}$. For the sake of simplicity we use a homogeneous conductor so that each filament of length ℓ has the same resistance R_f. If a time-varying current is driven along this conductor by an electromotive force e, which will be the same for every filament, then the current in the general filament m is governed by

$$e = i_m R_m + e_{b,m} \tag{3.87}$$

where $e_{b,m}$ is the back-e.m.f. induced in the filament m. The task facing us is to find this back-e.m.f. which depends on the rate of change of all of the filament currents and the mutual inductance between the filaments. Each filament current is controlled by an equation like

Eq.3.87. Unless the number of filaments is very small, the solution process is complex and time consuming. Not until computers became available was it possible to make any headway in this endeavour.

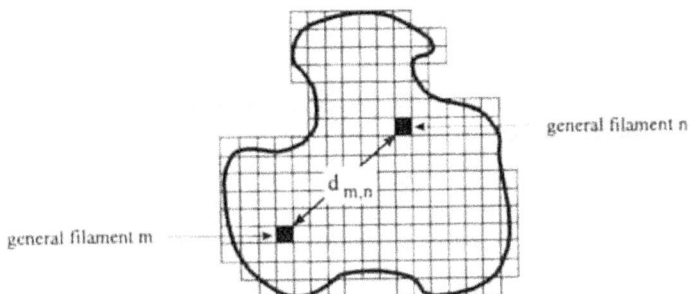

Figure 3.20 Conductor subdivision into square filaments

If we denote the mutual inductance between two general filaments by $M_{m,n}$ and the selfinductance of the general filament m by L_m, a set of simultaneous equations corresponding to Eq.3.87 may be structured as follows

$$e = (R_f i_1 + L_1 i_1') + M_{1,2} i_2' + \quad\cdots\quad + M_{1,g} i_g'$$

$$e = M_{2,1} i_1' + (R_f i_2 + L_2 i_2') + \quad\cdots\quad + M_{2,g} i_g'$$

$$e = M_{g,1} i_1' + M_{g,2} i_2' + \quad\cdots\quad + (R_f i_g + L_g i_g') \qquad (3.88)$$

where i_m' stands for the time derivative of i_m. When all currents are constant, Eq.3.88 reduces to the DC case in which the current distribution depends only on the resistances. When all filament resistances are the same the current distribution is uniform.

Let us immediately concentrate on the most important practical case in which the driving e.m.f., and therefore all filament currents, are sinusoidal of radian frequency ω, so that

$$M_{m,n} i_n' = j \omega M_{m,n} i_n = Z_{m,n} i_n \qquad (3.89)$$

$$R_f i_m + L_m i_m' = (R_f + j \omega L_m) i_m = Z_{m,m} i_m \qquad (3.90)$$

where $j = \sqrt{-1}$. $Z_{m,n}$ and $Z_{m,m}$ are now mutual and self impedances replacing the back-e.m.f.s of Eq.3.87.

The array of simultaneous equations Eq.3.88 may be abbreviated in matrix notation to

$$e = [Z] \{i\} \tag{3.91}$$

where $\{i\}$ is a vector or column matrix. The impedance matrix is square, symmetrical and of order g, the number of filaments in the conductor.

$$[Z] = \begin{array}{cccc} Z_{1,1} & Z_{1,2} & \cdots & Z_{1,g} \\ Z_{2,1} & Z_{2,2} & \cdots & Z_{2,g} \\ \\ \\ Z_{g,1} & Z_{g,2} & \cdots & Z_{g,g} \end{array} \tag{3.92}$$

One of the solutions of Eq.3.91 may be written

$$\{i\} = [Z]^{-1} e \tag{3.93}$$

Many sophisticated mathematical techniques have been developed to diminish the monumental amount of arithmetic involved in solving Eq.3.93. Silvester [3.9] was first to point out that the problem could be handled with what he called modal network theory. Not only does this method reduce the computational work, its main advantage is the ready availability of computer programs for determining the eigenvectors of any square symmetrical matrix like that in Eq.3.93.

For a small number of filaments the direct solution process, involving the determinant of the impedance matrix, Eq.3.92, may be employed. To illustrate this we take the simple example of the strip conductor of figure 3.21. The strip has been subdivided into three square-section (a × a) filaments. The length of the straight conductor is ℓ. If the strip is made of copper and a=1 cm and ℓ =1000 cm, then the room temperature resistance of each filament is $R_f = 1.76$ mΩ.

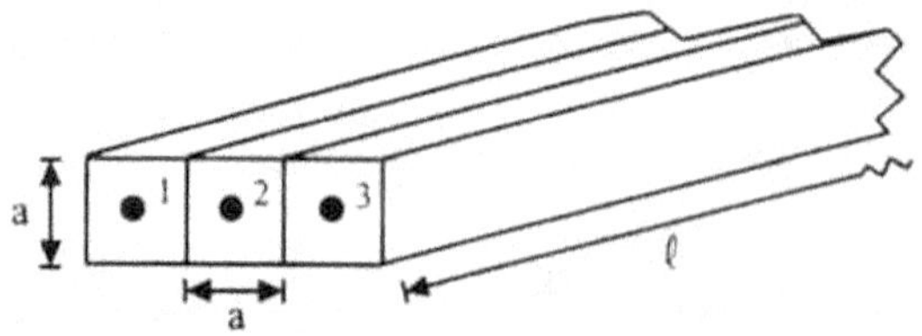

Figure 3.21 : Three filament rectangular strip conductor

Since the GMD of the square filament section is 0.44705a, the self and mutual inductances resulting from Eq.3.49 in SI units are

$$L_1 = L_2 = L_3 = 14\ 813\ \mu H$$
$$M_{1,2} = M_{2,3} = 13\ 204\ \mu H$$
$$M_{1,3} = 11\ 820\ \mu H$$

Let us write the simultaneous equations for the three-filament conductor as follows

$$Z_{1,1} i_1 + Z_{1,2} i_2 + Z_{1,3} i_3 = e$$
$$Z_{2,1} i_1 + Z_{2,2} i_2 + Z_{2,3} i_3 = e$$
$$Z_{3,1} i_1 + Z_{3,2} i_2 + Z_{3,3} i_3 = e \qquad (3.94)$$

where i_1, i_2, and i_3 are the three sinusoidal filament currents, e is the driving e.m.f. and the Z-impedances are complex. The selfinductances of the filaments are all the same, thus

$$Z_{1,1} = Z_{2,2} = Z_{3,3} = R_f + j\omega L_1 \qquad (3.95)$$

Also because $M_{1,2} = M_{2,3}$

$$Z_{1,2} = Z_{2,1} = Z_{2,3} = Z_{3,2} = j\omega M_{1,2} \qquad (3.96)$$

and

$$Z_{1,3} = Z_{3,1} = j\omega M_{1,3} \qquad (3.97)$$

The determinant D of the impedance matrix may be evaluated in terms of its cofactors. Then because of the equality of the selfinductances and certain mutual inductances, the determinant reduces to

$$D = Z_{1,1}^3 - 2 Z_{1,2}^2 (Z_{1,1} - Z_{1,3}) - Z_{1,1} Z_{1,3}^2 \qquad (3.98)$$

The numerical value of this determinant depends on the frequency of e. Let this be the power frequency f = 60 Hz. Then $\omega = 2\pi f = 377$ rad/s, and

$$Z_{1,1} = R_f + j\omega L_1 = (1.76 + 5.58 j) \times 10^{-3} \quad (\Omega)$$
$$Z_{1,2} = j\omega M_{1,2} = 4.98 \times 10^{-3} j \quad (\Omega)$$
$$Z_{1,3} = j\omega M_{1,3} = 4.46 \times 10^{-3} j \quad (\Omega)$$

and the determinant is found to be

$$D = (-36.64 + 44.66j) \times 10^{-9}$$

The three solutions of Eq.3.94 are given by

$$i_1 = \frac{C_1}{D}, \quad i_2 = \frac{C_2}{D}, \quad i_3 = \frac{C_3}{D} \tag{3.99}$$

C_1, C_2, and C_3 are modified impedance determinants determined by using Cramer's rule. From the symmetry of the conductor we know that $i_1 = i_3$ and therefore $C_1 = C_3$, thus only C_1 and C_2 need be computed

$$C_1 = e(Z_{1,1}^2 - Z_{1,1}Z_{1,2} + Z_{1,2}Z_{1,3} - Z_{1,1}Z_{1,3}) = e(2.43 + 3.03j) \times 10^{-6} \tag{3.100}$$

$$C_2 = e(Z_{1,1}^2 - 2Z_{1,1}Z_{1,2} + 2Z_{1,2}Z_{1,3} - Z_{1,3}^2) = e(3.01 + 2.12j) \times 10^{-6} \tag{3.101}$$

From this it follows that

$$i_1 = i_3 = (C_1/D) = (13.87 - 65.79j)e$$
$$i_2 = (C_2/D) = (-4.68 - 63.56j)e$$

The amplitude and phase relationships of the filament currents with respect to the driving e.m.f. are shown in figure 3.22. As expected, the current amplitudes in the two outer filaments are greater than in the center filament. This is a manifestation of the skin effect. It would be more pronounced at higher frequencies. Figure 3.22 also shows the phase differences between the filament currents. The total current I flowing in the conductor is the phasor sum of the filament currents, or

$$I = 2(13.87 - 65.79j)e + (-4.68 - 63.56j)e$$
$$= (23.06 - 195.14j)e$$
$$|I| = 196.5e \quad \text{(lagging by } 83.3°)$$

This example demonstrates that isolated linear conductors made of copper or aluminum are highly inductive, with the AC current lagging the driving e.m.f. by nearly ninety degrees. When two closely spaced conductors form a go-and-return circuit, however, their combined inductance is greatly reduced.

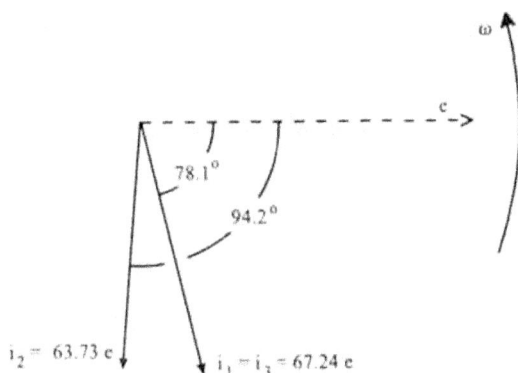

Figure 3.22 : Phasor diagram of the filamentary currents in the conductor shown in figure 3.21

One of the primary reasons for computing the AC current distribution is to determine what is known as the AC-resistance. In the present example this would be given by

$$ R_{ac} = \left(\frac{2|i_1|^2 + |i_2|^2}{|I|^2} \right) R_f = 0.3394 \ R_f $$

The DC-resistance of the strip conductor is $0.3333 \ R_f$. Hence by passing 60 Hz current, instead of DC, through the conductor the Joule heating is increased by just under two percent. At higher frequencies this increase would of course be greater.

Finally we note that the AC current distribution is a steady state distribution in which the current in the center is permanently smaller than at the perimeter. This will not be the case for pulse currents. At the pulse front the current distribution is similar to the AC case, but on the decaying side of the pulse the current is more concentrated in the core of the conductor than on the outside. This is a little known phenomenon which may be called the 'core effect' [3.10].

A qualitative explanation of the core effect can be obtained from the back-e.m.f. equation, Eq.3.87. On the rising side of the pulse, where de/dt and all di/dt are positive, the back-e.m.f. will oppose current flow. It is a maximum where the inductance is greatest, that is where a filament has the most close neighbors. Hence on the pulse front the back-e.m.f. reduces current flow in the center of the conductor. At the peak of the pulse, when de/dt and all di/dt are zero, the current distribution will be governed by the filament resistances. Now on the decaying side of the pulse the time derivatives of e and i are everywhere negative, and the back-e.m.f.s reverse sign and assist continued current flow, mostly in the center of the conductor and least at the perimeter.

A quantitative proof of the core effect may be provided with the help of Laplace transforms. They require the conversion of a function of time f(t) to the function f(s) of the Laplace operator s by

$$f(s) = \int_0^\infty f(t)\, e^{-st}\, dt \qquad (3.102)$$

Algebraic operations are then performed on f(s) until it is in a convenient form for re-conversion to the desired solution f(t). Even for the simple three-filament conductor of figure 3.21 the procedure is quite lengthy. An outline will be given here which omits much of the detail. An excellent account of the handling of such transient currents by Laplace transforms has been published by Greenwood [3.11].

The starting point is to set up the array of three simultaneous equations for the conductor of figure 3.21. The format of this array has been provided by Eq.3.88. Instead of applying a voltage pulse we will consider separately the switch-on transient e = constant for t > 0, and then later the switch-off transient e = 0 for t > 0. The combination of the two is equivalent to a flat top voltage pulse. In the case of the switch-on transient the initial current in each filament is zero and the Laplace transforms of the filament current and its time derivative are simply I and I.s, respectively. Furthermore, the Laplace transform of the voltage step e is e/s. With this notation, and remembering that all three filaments have the same resistance R and selfinductance L, and $M_{1,2} = M_{2,3}$, the three simultaneous equations may be written

$$s(R+Ls)I_1 + M_{1,2}s^2I_2 + M_{1,3}s^2I_3 = e$$
$$M_{1,2}s^2I_1 + s(R+Ls)I_2 + M_{1,2}s^2I_3 = e$$
$$M_{1,3}s^2I_1 + M_{1,2}s^2I_2 + s(R+Ls)I_3 = e \qquad (3.103)$$

With the resistance and inductance values which have already been specified for the AC example, the impedance determinant of Eq.3.103 can be shown to be equal to

$$D = 1.918 \times 10^{-16}s^3(s+44.19)(s+588.09)(s+1093.7) \qquad (3.104)$$

The two modified impedance determinants corresponding to Eqs.3.100 and 3.101 are found from

$$C_1 = \begin{vmatrix} e & M_{1,2}s^2 & M_{1,3}s^2 \\ e & s(R+Ls) & M_{1,2}s^2 \\ e & M_{1,2}s^2 & s(R+Ls) \end{vmatrix} \qquad (3.105)$$

$$C_2 = \begin{vmatrix} s(R + Ls) & e & M_{1,3}s^2 \\ M_{1,2}s^2 & e & M_{1,2}s^2 \\ M_{1,3}s^2 & e & s(R + Ls) \end{vmatrix}$$

(3.106)

After the appropriate numerical substitutions, these last two determinants are found to be equal to

$$C_1 = 4.816 \times 10^{-12} s^2 (s^2 + 1682 s + 6.432 \times 10^5)e$$

(3.107)

$$C_2 = 6.734 \times 10^{-13} s^2 (s^2 + 8410 s + 4.600 \times 10^6)e$$

(3.108)

Hence the Laplace transform of the currents in filaments 1 and 3 is

$$I_1 = I_3 = \frac{C_1}{D} = \frac{25109 \, e(s^2 + 1682 \, s + 6.432 \times 10^5)}{s(s + 44.19)(s + 588.09)(s + 1093)}$$

(3.109)

Similarly, I_2, the current in the center filament is

$$I_2 = \frac{C_2}{D} = \frac{2.043 \times 10^6 e(s^2 + 8410 \, s + 4.6 \times 10^6)}{s(s + 44.19)(s + 588.09)(s + 1093.7)}$$

(3.110)

Equations 3.109 and 3.110 have to be split into partial fractions for which the inverse Laplace transforms are known. This can be achieved with

$$\frac{s^2 + us + v}{s(s - a)(s - b)(s - c)} = \frac{A_1}{s(s - a)} + \frac{B_1}{(s - a)(s - b)} + \frac{C_1}{(s - b)(s - c)}$$

(3.111)

and the identity

$$s^2 + us + v = A_1(s - b)(s - c) + B_1 s(s - c) + C_1 s(s - a)$$

(3.112)

The three partial fraction numerators come to

$$A_1 = \frac{v}{bc} \tag{3.113}$$

$$B_1 = \frac{a + u + \dfrac{v}{a}}{a - c} - \frac{A_1(a - b)}{a} \tag{3.114}$$

$$C_1 = \frac{c + u + \dfrac{v}{c}}{c - a} \tag{3.115}$$

It follows from Eqs.3.109 and 3.111 that the inverse transform of the currents in filaments 1 and 3 may now be written

$$i_1 = i_3 = 25109 \; e\left(\frac{A_1(1 - \epsilon^{-at})}{-a} + \frac{B_1(\epsilon^{-at} - \epsilon^{-bt})}{a - b} + \frac{C_1(\epsilon^{-bt} - \epsilon^{-ct})}{b - c} \right) \tag{3.116}$$

where ϵ represents the base of natural logarithms, used here to distinguish it from the e.m.f., e. Using Eq.3.112, another set of partial fraction numerators A_2, B_2, and C_2 may be found for the current in the middle filament expressed in Eq.3.110. The time dependence of this current may then be expressed by

$$i_2 = 2.043 \times 10^6 \; e\left(\frac{A_2(1 - \epsilon^{-at})}{-a} + \frac{B_2(\epsilon^{-at} - \epsilon^{-bt})}{a - b} + \frac{C_2(\epsilon^{-bt} - \epsilon^{-ct})}{b - c} \right) \tag{3.117}$$

It should be noted that the roots a, b, and c are the same for i_1 and i_2 and they are all negative. Substituting the previously specified filament resistances and self and mutual inductances for obtaining the A, B and C coefficients, evaluation of the filament currents by Eqs.3.116 and 3.117 resulted in the figures listed in table 3.6 which apply to the conductor of figure 3.21.

One second after applying the electromotive force to the conductor, the currents have attained their steady state value of 568.2 A. This is equal to e/R when e = 1 volt. It will be seen that during the switch-on transient the current in the center filament is smaller than the current in the two outer filaments. In this particular calculation the difference between i_1 and i_2 is quite small. The true current distribution would be less uniform because the small number of three filaments gives too crude an approximation. This approximation hides the fact that the current density will be a maximum at the corners of the conductor section where a thin filament has fewer neighbors than any other in the conductor section. It is the multiplicity of near-neighbor filaments which delays current growth in the center of the conductor. Computer solutions for over 100 filaments are feasible and they would show the details of the current distribution quite accurately.

The switch-off transient current distribution has been calculated by the same method and the results are also shown in table 3.6. They demonstrate the core-effect.

	Switch-on Transient			Switch-off Transient	
t(s)	i_1 (A)	i_2 (A)	t(s)	i_1 (A)	i_2 (A)
0.000	0.0	0.0	0.000	568.2	568.2
0.001	24.6	17.9	0.001	493.0	859.3
0.005	112.7	97.6	0.002	507.3	720.2
0.006	132.3	117.4	0.003	495.3	623.7
0.007	151.2	136.5	0.004	479.8	557.4
0.008	169.2	155.0	0.005	460.4	510.5
0.009	186.5	172.8	0.006	440.9	475.5
0.010	203.0	189.8	0.007	422.0	447.6
0.020	333.4	324.9	0.008	403.8	424.2
0.030	417.3	411.8	0.009	386.4	403.5
0.040	471.2	467.7	0.010	369.7	384.8
0.050	505.8	503.6	0.015	296.4	307.2
0.060	528.1	526.7	0.020	237.7	246.3
0.070	542.4	541.5	0.025	190.6	197.4
0.080	551.6	551.1	0.030	152.8	158.3
0.090	557.6	557.2	0.035	122.5	127.0
0.100	561.4	561.2	0.040	98.2	101.8
0.110	563.8	563.7	0.045	78.8	81.6
0.120	565.4	565.3	0.050	63.1	65.4
0.130	566.4	566.4	0.055	50.6	52.4
0.140	567.0	567.0	0.060	40.6	42.0
1.000	568.2	568.2	0.065	32.6	33.7
			0.080	16.8	17.4
			0.100	6.9	7.2

Table 3.6 : Current distribution in strip conductor shown in figure 3.21

Field theory relies on the idea of magnetic field diffusion into the conductor metal for explaining the skin effect. The diffusion model cannot predict the core effect. Presumably, this is the reason why there is no mention in the literature of switch-off transient current distributions.

A rigorous solution of magnetic field diffusion exists only for infinite half-space. The resulting skin-depth formula is a poor approximation for wires and cables. Maxwell [1.8] fell back on the Ampère-Neumann filament model in developing his elliptic integral technique for calculating the skin depth in round conductors. Not until computer assisted finite filament analysis became available was it possible to investigate the AC current distribution in rectangular conductors.

We may consider the two switching transients to be the front and back of a square voltage pulse. At the front the e.m.f. applied to each filament is the same. This exerts a constraint on the current distribution which is absent in the back of the pulse. When the external e.m.f. is removed, the induced e.m.f.s in the filaments may differ, causing a lateral current exchange. This is the reason why during the switch-off transient i_2 is initially greater than 568.2 A, the steady state current in the conductor. After about four milliseconds the lateral exchange of current appears to have died out.

When the number of filaments is greater than ten it becomes advisable to abandon the foregoing solution process. Kelly [3.12] has outlined the best-known alternative techniques suitable for computers. As previously mentioned, Silvester [3.9] found that modal network analysis provides additional tools for solving the set of simultaneous linear equations. This involves eigenvectors (mode vectors) and eigenvalues (mode frequencies).

The Induction of Eddy Currents

The term "eddy currents" is a misnomer. It suggests flow irregularities when, in fact, electromagnetic induction is very precise and orderly. Nevertheless, the subject is complex and continuously spawns new publications in an already vast literature. In several instances the Newtonian electrodynamics has proved to be a more powerful tool than field theory for solving eddy current problems. But there is no room in this book for the voluminous analysis of induced currents in three-dimensional conductors. Instead the reader will be referred to some of the relevant publications.

Eddy currents are of practical importance in induction heating, non-destructive testing, shielding, and in a variety of AC power problems. In the 1950s one of the authors (PG) became involved in what was then known as eddy current testing of travelling wires and metallic pipes. The relative motion between test object and detector coils led to dynamic induction which was difficult to visualize in field theory, but became more apparent in Neumann's theory [3.13, 3.14, 3.15].

Two significant facts emerged from the early investigation which ultimately became responsible for this book. The first was that a Newtonian action at a distance theory could explain precisely the same facts, related to relative motion, as electromagnetic field theory with Einstein's special relativity. The second fact concerned the time delay between the cause of induction and the induced effect itself. This time delay, or the corresponding AC phase shift, could be explained with equal precision by the energy transport time lag of field theory or the many-body simultaneous matter interaction process of the Newtonian electrodynamics. This second fact suggests that a time may come when the eight minutes it takes sunlight to reach the earth can be accounted for by a simultaneous far-action theory.

A Neumann induction model was developed which involved sequential filament interactions. It resulted in converging infinite series solutions of the eddy current distribution. The terms of the series contained successively higher-order time derivatives of the cause of the induction. These series produced the phase shifts corresponding to energy transport delays in field theory. In terms of physics, the model suggests that when nature is faced with the inevitable complexity of multiple interactions (the many-body problem) it adjusts to the final state by a series of steps which may be infinitely small and infinite in number. Each step takes the result of the previous step into account in the long chain of interactions that all occur simultaneously. It could be the way nature arrives at eigenvalue solutions.

Chapter 3 References

3.1   F.W. Grover, *Inductance calculations*, Dover, New York, 1945.

3.2   E.D.Charles,"Mechanical forces on current-carrying conductors", Proceedings IEE, Vol.110, p.1671, 1963.

3.3   F.F. Cleveland, "Magnetic forces in a rectangular circuit", Philosophical Magazine, Vol.21, p.416, 1936.

3.4   W. Thompson (Lord Kelvin), *Mathematical and physical papers*, Vol.1, p.521, 1853.

3.5   A. Sommerfeld, *Electrodynamics*, Academic Press, New York, 1952.

3.6   P. Graneau, "A re-examination of the relationship between self and mutual inductance", Journal of Electronics and Control, Vol.12, p.125, 1962.

3.7   P. Graneau, *Underground power transmission*, Wiley, New York, 1979.

3.8   Lord Rayleigh, "On the selfinduction and resistance of straight conductors", Philosophical Magazine, Vol.21, p.381, 1886.

3.9   P. Silvester, "Modal network theory of skin effect in flat conductors", Proceedings IEE, Vol.54, p.1147, 1966.

3.10   P. Graneau, "Alternating and transient conduction currents in straight conductors of any cross-section", International Journal of Electronics, Vol.19, p.41, 1965.

3.11   A. Greenwood, *Electrical transients in power systems*, Wiley, New York, 1971.

3.12   L.G. Kelly, *Handbook of numerical methods and applications*, Addison-Wesley, Reading MA, 1967.

3.13   P. Graneau, "Coupled circuit theory for electromagnetic testing", *Progress in non-destructive testing*, (E.G. Stanford, Editor), Heywood, London, Vol.3, p.163, 1961.

3.14   P. Graneau, "Coupled circuit theory for electromagnetic testing", Ph.D. Thesis, University of Nottingham, U.K., 1962.

3.15   P. Graneau, "Steady-state electrodynamics of a cylindrical body in axial motion", Journal of Electronics and Control, Vol.14, p.459, 1963.

CHAPTER 4

The Nature of Current Elements

Current Elements and Newton's Third Law

This chapter deals with the most fundamental aspect of electrodynamics. Since the question of what constitutes a current element is far from settled, Chapter 4 is inevitably more speculative than the remainder of the book. The experimental evidence which was presented in Chapter 2 leaves no doubt that the ponderomotive electrodynamic forces -- but not the electromotive forces -- act directly on the metal lattice. Hence the Ampèrian current element must be something very different than the drifting conduction electron which is mechanically decoupled from the lattice ions. There seems to exist no rational grounds for the claim of conventional electromagnetic field theory that one and the same force law should apply to both the flow of electric current in a wire and an electron beam in a cathode ray tube. If more than one force law is required it would not be surprising that more than one type of current element must be considered.

Ampère wrote the first chapter in the current element story. To him one of these elements was simply an infinitely small piece of conductor which contained an infinitely small amount of electric fluid in motion relative to the metal. When the motion of the fluid stopped, the element became inactive. This was the *particle of electrodynamics* on which a theory could be based which had Newton's gravitation as its model. Maxwell [1.8] studied Ampère's work more closely than any other scientist. In a recapitulation of his chapter on the principles of electrodynamics, Maxwell reminded his readers:

> "It must be carefully remembered, that the mechanical force which urges a conductor carrying a current across the lines of magnetic force, acts, not on the electric current, but on the conductor which carries it. .... The only force which acts on the electric current is the electromotive force, which must be distinguished from the mechanical force which is the subject of this chapter."

As the clear distinction between ponderomotive and electromotive forces was obliterated by the introduction of the Lorentz force law, long after Maxwell's death, and since in Chapter 2 this modern law of relativistic electromagnetism was shown to be flawed, we will quote Maxwell [1.8] once more when he said:

> "Electromotive force is always to be understood to act on electricity only, not on the bodies in which the electricity resides. It is never to be confounded with ordinary mechanical force, which acts on bodies only, not on the electricity in them."

In his development of the early stages of a Newtonian electrodynamics, Ampère restricted himself to the ponderomotive interaction of current elements. It was Franz Neumann who subsequently addressed electromotive forces.

According to a translation by Tricker [1.5] Ampère said:

> "Newton taught us that motion of this kind, like all motions in nature, must be reducible by calculation to forces acting between two material particles along the straight line between them such that the action of one upon the other is equal and opposite to that which the latter has upon the former and, consequently, assuming the two particles to be permanently associated (rigidly interconnected) that no motion whatsoever can result from their interaction."

Apart from the fact that Ampère's statement does not mention the simultaneity of the two-body interaction, this is one of the most concise formulations of Newton's third law. Short forms of the law like 'action and reaction are equal and opposite' are incomplete and can be misleading.

Newton's own formulation of the third law reads as follows:

> "To every action there is always opposed an equal reaction: or the mutual actions of the two bodies upon each other are always equal and directed to contrary parts."

The important aspects of Newton's third law may be summarized as follows:

(1)   The law refers to a mutual ponderomotive interaction force between no more and no less than two material entities.

(2)   It does not deal with the interaction between matter and a non-material medium like ether or field energy.

(3)   The interaction force is a single force of attraction or repulsion.

(4)   No messages are exchanged between the interacting matter entities. Since the magnitude of the attraction or repulsion depends on the distance between, and the properties of, the two matter elements, both particles must at all times and simultaneously be aware of their properties and distance of separation.

(5)   The third law does not apply to electromotive forces which are non-reciprocal forces.

Ampèrian Current and Conductor Elements

There was no doubt in the minds of Ampère and his followers that the forces defined by Ampère's law, Eq.1.24, acted directly on the substance of the conductor and not on the electric fluid which was free to slide through the metal without drag or any other mechanical interaction. Ampère also knew that in order to calculate the force between two circuits, the current elements did not have to be infinitely small, so long as the separation between them was large compared to their size. Weber would later speak of Ampère's ponderable wire elements. Precisely these wire elements have been resurrected in computer assisted finite current element analysis. Because of the absence of mechanical linkage between the electric fluid and the conductor metal, Ampère's electrodynamics has survived such innovations as atomicity, crystallinity, electron structure, and quantum mechanics.

In the twentieth century the electric fluid of the old electrodynamics has become the Fermi-sea of conduction electrons. These electrons collide with lattice ions and crystal imperfections and thereby generate Joule heat. But the mechanical forces arising in electron-lattice collisions are negligibly small and are certainly unable to account for the longitudinal forces predicted by Ampère's law. The only way a conduction electron can exert a significant force on the body of the metal is by running up against the surface barrier (work function) when Coulomb forces between the electron and metal ions come into play. Hence the longitudinal electrodynamic forces of Chapter 2 must be forces between lattice ions and their atomic structures. This means, ultimately, that Ampèrian current elements have to be atoms which have become magnetically polarized owing to the flow of electrons. At the time of writing, the magnetic make-up of the Ampèrian current element remains unknown. Present speculations point to a diamagnetic dipole. It is certain, however, that the current element cannot be infinitely small.

A postulated property of any Newtonian matter element is that it will not interact with itself. This applies to atoms, current elements, and any particle which can be described as fundamental and indivisible. Finite element analysis has proved that this postulate of no self-interaction may also be applied to macroscopic current elements which are clusters of atomic current elements. Because of the absence of self-interactions, Ampèrian current elements are invariably studied in pair combinations. All actions of the Newtonian electrodynamics are mutual actions between the members of a pair of elements. For mutual mechanical forces to exist, both elements must carry currents.

In the case of electromagnetic induction, however, just one of the two elements need carry a current; the other can be a neutral element of conductor metal or a neutral atom. There exists no reaction force to the induced electromotive force. Since this force does not act on the ponderable substance of the conductor, and is not a force measured in newtons, it need not comply with Newton's third law.

Although he lived through the second half of the nineteenth century, Neumann remained as vague as Ampère about the inevitable atomic nature of the current and conductor elements. To this day, the same vagueness can be found in many modern texts on electromagnetism which, by their silence on the make-up of the metallic current element, perpetuate the antiquated notion of the infinitely divisible subtile fluid of electricity. Neumann's extension of the Ampère theory dealt with the electromotive forces on the imponderable electric fluid. The parallel existence of both ponderomotive and electromotive forces has become the hallmark of Newtonian electromagnetism.

Weber's Current Elements

The first step toward the development of the atomic current element concept was taken by Weber [1.17]. His metallic elements are depicted in fig.1.18. He conjectured positively and negatively charged particles moving through the metal at equal velocity in opposite directions. His force law, Eq.1.61, which agrees with Ampère's law, Eq.1.24, involved equal and opposite charge velocities, with respect to the conductor metal, in terms of dr/dt derivatives. Weber pointed out that in his current element model the charge velocity v could never become greater than c, which we now know as the velocity of light. This suggests that Weber's theory already contained the essence of Einstein's special relativity. Furthermore, this state of affairs arose from attributing magnetic actions to the finite velocity of charged particles, an idea which later played a dominant role in Maxwell's theory. It introduced the dimension of time into what was originally a simultaneous far-action law. The ponderomotive force laws of Newton, Coulomb, and Ampère did not contain time.

When there is no current flowing in the metal, and v = 0 in figure 1.18, the charges in the Weber current element are still subject to Coulomb forces. However these interactions cancel because each current element is said to be *charge neutral*. This avoids a clash between time-dependent and time-independent forces. The two charges of opposite polarity in the same Weber current element are assumed not to attract each other, and yet stick together, thus defying Coulomb's law. In a similar vein, it should be recalled that the defiance of Coulomb's law inside the atom led to the creation of quantum mechanics.

Although a strong advocate of Coulomb's law, Weber did not suggest that the two charges of his 'neutral' element would combine, for in that case electromotive forces would not be able to start current flow in metal wires. Even though there were logical problems, the invention of a current element which combined Coulomb and Ampère forces was truly remarkable.

Today we know that the positive charges in metal are fixed to the lattice and cannot move relative to it. Hence the model depicted in figure 1.18 has to be modified so that only the negative charges travel relative to the conductor. Recently Assis [1.19] has proved that, in spite of this modification, the Weber electrodynamic force remains mathematically equivalent to Ampère's force law. It has to be remembered, however, that this mathematical proof is valid only if the stationary Ampèrian atomic current element is abandoned and replaced by a travelling negative charge and a fixed positive charge.

With the positive charge fixed to the metal lattice, Weber's force law is now clearly able to impart some ponderomotive action to the body of the metal. However for complete physical compliance of Eq.1.61 with Ampère's law, the negative conduction electrons must also exert mechanical forces on the lattice. This appears to be impossible unless the free mobility of the conduction electrons, prescribed by the energy band model of quantum theory, is abandoned, and with it our whole understanding of the electrical conduction process.

Weber's electrodynamics involves two quite different current elements, one for metals and the other for plasma. The metal element is shown in figure 1.18. With the positive charge at rest relative to the conductor, this is Weber's updated current element, still exerting force on the metal. When electric current flows through a plasma, or is convected by charges through electrolytes or vacuum spaces, the Weber current element becomes simply a travelling charge for there then exists no conductor. The relative velocity is no longer the velocity relative to a piece of metal, but must be the relative velocity between the interacting charges.

In an electron beam the relative velocity may be different for a pair of adjacent electrons and another widely separated pair. Yet the electron beams of cathode ray tubes are remarkably stable. It would be of interest to know if this stability can be explained with Weber's force law and his current element model of the travelling charge. An analysis of this problem has not been published. If charges drifting in vacuum really obey Eq.1.61, as Assis [1.19] and Wesley [4.1] claim, then the Newtonian electrodynamics could be extended to particle physics which, so far, has remained the domain of field theory and special relativity.

In Wesley's view [4.1], Weber came close to the quantization of charge. Weber assumed the positive and negative charges in his metallic current element to be equal in magnitude to each other. This is certainly true for the conduction electron and its associated lattice ion, even though in the 1840s, Weber had no knowledge of the electron theory of metals. With neither of his charges being mechanically coupled to the substance of the conductor, he painted a confused picture of how forces were transferred from the charges to the body of the metal. At the same time the mobile charges could respond well to electromotive forces. He considered this to be an advantage of his current element over that of Ampère.

The Modern Current Element

The essential modification which Lorentz [4.2] made to the metallic current element of Weber was to drop the positive charge and eliminate the containment box in figure 1.18 which would surround the remaining negative charge. This box was meant to represent the mechanical linkage between the charges and the metal lattice. The removal of the box presents the question of how the conduction electron interacts with the metal atoms ?

Consider the Lorentz force $\vec{F}$ defined by

$$\vec{F} = q\ (\vec{E} + \vec{v} \times \vec{B})\ \ (4.1)$$

where q is the electric charge which experiences the force and $\vec{E}$ and $\vec{B}$ are the electric field strength and the magnetic induction vectors, respectively, at the location of q. The velocity $\vec{v}$ is the velocity of the charge relative to an arbitrary observer. The current element is $q\vec{v}$. Lorentz ruled that the force on a current element, $\vec{F}$, must always be perpendicular to the direction of current in the element. He considered this to be an empirical fact with the consequence that no force was exerted on the metal lattice in the direction of current flow. The empirical knowledge leading to his reasoning was, what he considered to be, the absence of observed longitudinal electrodynamic forces. This enabled him to remove Weber's box around the metallic current element. It turned out to be a grave error which diverted electrodynamics away from the Newtonian school of thinking.

Another -- this time valid -- empirical fact was that continuous currents in metals always formed closed loops which did not intersect any surfaces of the conductor. Electrons would, however, meet surface barriers in the direction transverse to $\vec{v}$ and then could exert (electrostatic) force on the metal. This was Lorentz's second empirical fact justifying his force law.

The modern current element is taken to be the product of any electric charge, positive or negative, multiplied by its velocity $\vec{v}$ relative to an arbitrary observer. If we think in terms of the interaction of two travelling charges, we might mistakenly take $\vec{v}$ to be the relative velocity between the charges. This would not comply with the experimental facts of electromagnetic induction and in fact led directly to the formulation of the special theory of relativity. In the first paragraph of the special relativity paper [4.3], which Einstein entitled 'On the electrodynamics of moving bodies', he discussed the effect of relative motion between a permanent magnet M and a conducting body C, as shown in figure 4.1. If $\vec{v}_m$ and $\vec{v}_c$ are the velocities of the magnet and the conductor relative to the laboratory, then the relative velocity between the two objects is

$$\vec{v}_r = \vec{v}_m - \vec{v}_c \qquad (4.2)$$

So long as the relative velocity is not zero, there will be currents induced in the conductor which can be measured with the ammeter A of figure 4.1. Regarding this experimental fact Einstein wrote:

"It is known that Maxwell's electrodynamics -- as usually understood at the present time -- when applied to moving bodies, leads to asymmetries which do not appear to be inherent in the phenomena. Take, for example, the reciprocal electrodynamic action of a magnet and a conductor. The observable phenomenon here depends only on the relative motion of the conductor and the magnet, whereas the customary view draws a sharp distinction between the two cases in which either the one or the other of these bodies is in motion. For if the magnet is in motion and the conductor at rest, there arises in the neighborhood of the magnet an electric field with a certain definite energy, producing a current at the places where parts of the conductor are situated. But if the magnet is stationary and the conductor in motion, no electric field arises in the neighborhood of the magnet. In the conductor, however, we find an electromotive force, to which in itself there is no corresponding energy, but which gives rise -- assuming equality of relative motion in the two cases discussed -- to electric currents of the same path and intensity as those produced by the electric forces of the former case."

Thus in special relativity, if we wish to read the ammeter in the normal way we would have to move with the conductor C. Then we would confirm the existence of electromagnetic induction. The implication is that an observer who is stationary in the laboratory cannot read the instrument and therefore, may draw wrong conclusions about what causes induction. This is the essence of the observer-physics introduced by Einstein, and developed with the aid of the Lorentz transformations. It is however incorrect, for we could set up a video-camera in the laboratory which would record the meter deflection when the ammeter flies by the lens. The conclusion we have to draw from this is that Maxwell's equations are at fault. An experiment will be described in Chapter 5 which reveals that Maxwell's equations cannot deal satisfactorily with all situations of electromagnetic induction.

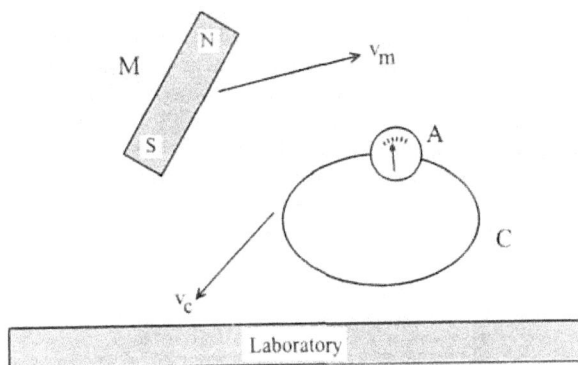

Figure 4.1 : Einstein's example of relative motion between a permanent magnet M and a conductor C

Mutual Torques between Ampèrian Current Elements

The mutually stored magnetic energy of two closed circuits m and n is given by Neumann's electrodynamic potential of Eq.1.35. The mutually stored energy of two current elements $i_m dm$ and $i_n dn$ is not necessarily the integrand of this equation because in closed-loop integrations we may have ignored energy contributions which are total differentials of the function which is being integrated. This fact is normally described by the notion of gauge invariance.

The well established fact that the reaction forces between two parts of the same current-carrying circuit depend on the selfinductance gradient was expressed by Eq.3.16. It forced us to come to terms with what was meant by selfinductance. Then it was found, in the Newtonian electrodynamics, that this was actually the mutual inductance of all conductor element combinations of the circuit. Each mutual inductance of a pair of conductor elements is given by Eq.3.19. The first treatment of this subject was published in 1985 [1.12].

From Neumann's Eq.1.35 and the mutual inductance of a pair of conductor elements, Eq.3.19, it follows that the mutually stored energy of two elements $i_m dm$ and $i_n dn$ is given by

$$\Delta P_{m,n} = - i_m i_n \frac{dm \ dn}{r_{m,n}} (2 \cos \varepsilon - 3 \cos \alpha \cos \beta) \qquad (4.3)$$

Since the stored energy $\Delta P_{m,n}$ changes with the angles of Eq.4.3, there must exist torques which represent the release of stored energy. The elemental electrodynamic torques were only discovered in 1985 [1.12], and the evolving theory, not surprisingly, has caused confusion. It is almost inevitable that errors may have to be corrected later when the elemental torques become better understood.

Equation 4.3 contains three angles. Since the calculation of the torques involves partial differentiation with respect to the angles, it will lead to the sines of the angles in the expression. As a result a slightly more precise angle convention than that used for the force calculations must be applied, which is described by figure 4.2. In addition ε is not an independent variable, and thus must be removed by the following substitution.

$$\varepsilon = \beta - \alpha \qquad (4.4)$$

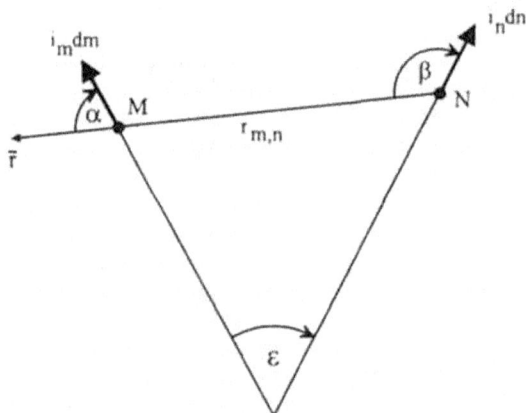

Figure 4.2 : Angle convention for torque calculations

The mutual torque between the two general current elements will be denoted by $\Delta T_{m,n}$. Rotation of either α or β or both can lead to both positive and negative values of $\Delta P_{m,n}$. Thus it appears ambiguous whether the elements will tend to a negative, positive or zero value for the stored energy. The empirical result, ususally referred to as Lenz's law implies that circuits induce currents in other circuits which lead to repulsive forces. Repulsive forces are represented by a positive mutual inductance. If we assume that the same principle applies to the interaction between two elements, then Neumann's virtual work concept predicts the following two mutual torques

$$(\Delta T_{m,n})_\alpha = \frac{\partial}{\partial \alpha} \Delta P_{m,n} \tag{4.5}$$

$$(\Delta T_{m,n})_\beta = \frac{\partial}{\partial \beta} \Delta P_{m,n} \tag{4.6}$$

Substituting Eqs.4.3, 4.4 and 4.5 gives

$$(\Delta T_{m,n})_\alpha = - i_m i_n \frac{dm \ dn}{r_{m,n}} \frac{\partial}{\partial \alpha} (2 \cos (\beta - \alpha) - 3 \cos \alpha \cos \beta) \tag{4.7}$$

After performing the partial differentiation it is found that

$$(\Delta T_{m,n})_\alpha = - i_m i_n \frac{dm \ dn}{r_{m,n}} (\sin \alpha \cos \beta + 2 \sin \beta \cos \alpha) \tag{4.8}$$

A positive alpha-torque tends to increase the angle α.
    Similarly, for the beta-torque of Eq.4.6 we can write

$$(\Delta T_{m,n})_\beta = - i_m i_n \frac{dm \ dn}{r_{m,n}} \frac{\partial}{\partial \beta} (2 \cos (\beta - \alpha) - 3 \cos \alpha \cos \beta) \tag{4.9}$$

And differentiating Eq.4.9 gives

$$(\Delta T_{m,n})_\beta = - i_m i_n \frac{dm \ dn}{r_{m,n}} (\cos \alpha \sin \beta + 2 \cos \beta \sin \alpha) \tag{4.10}$$

A positive beta-torque tends to increase the angle β.
    A torque $\Delta T_{m,n}$ has to be treated as a mutual torque which affects both elements $i_m dm$ and $i_n dn$. However as indicated in figure 4.3, the alpha torque, $(\Delta T_{m,n})_\alpha$ for instance, has the effect of trying to change the direction of $i_m dm$, while at the same time changing the position of $i_n dn$. Similarly the beta torque, $(\Delta T_{m,n})_\beta$, attempts to change the direction of $i_n dn$ and move the position of $i_m dm$. In other words, each torque has both an electromotive and a ponderomotive component.
    Whether it is possible for the current element to turn freely about the vector pivot depends on the, as yet unknown, microscopic structure of the element. In previous chapters it has been shown that the fundamental Ampèrian current element cannot be infinitely small. The smallest size appears to be the current-carrying conductor atom, or possibly the nucleus.

In the quantum mechanical theory of metallic conduction, the charge carrier, that is the conduction electron, does not exert ponderomotive forces on the lattice. It therefore has to be the atom which experiences the mechanical force given by Ampère's law. In order to have a direction, this atomic element would have to be a dipole of some form. As atoms, or nuclei, are not restrained from turning in the metal lattice, it will be assumed that the alpha and beta torques can indeed turn the dipole element about the center of the atom.

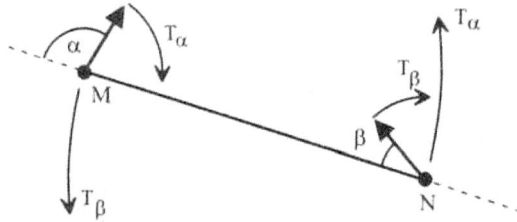

Figure 4 3   Effects of positive α and β torques

In a solid lattice under low current conditions, the ponderomotive components of the torques are assumed to be taken up by the lattice, and thus the torques will have a purely electromotive effect. In this situation, any pair of elements will attempt to rotate to their stable relative positions, which is when they are both pointing in opposite directions and perpendicular to the distance vector between them. This stable arrangement is shown in figure 4.4, and represents the effect of Lenz's law in which the two elements align themselves into a position where they experience maximum repulsion.

Figure 4.4   Stable position for two elements with zero mutual torque

A two-dimensional study of the interaction of up to 400 co-planar elements has revealed that the torques oppose the existence of stable parallel current filaments formed by in line elements, which must exist when current is flowing through a conductor. It seems therefore that the electromotive force responsible for current flow must provide energy to counteract the torques and allow current filaments to be created and maintained.

The effect of the ponderomotive torque components remains hidden in a metal lattice that can provide reaction forces to them as long as it does not break. However the effect of these torques in a plasma or liquid metal conductor remains unexplored. It seems likely that the two ponderomotive torque components on a pair of parallel elements will act in such directions as to expand the current carrying cross-section. These forces have to be taken into consideration along with the attraction and repulsion forces discussed in Chapter 2. The full theoretical prediction is not yet complete, and no experiments have yet been performed to check these predictions. Nevertheless the fact that arc columns are known to expand is consistent with the notion of the ponderomotive torque components. In Chapter 6, the issue of radial arc expansion will be discussed in more detail.

In our model of the Newtonian electrodynamics it rather looks as if the magnetic effects of the current are due to the alignment of dipole elements. The model implies that the dipoles exist even when there is no current flowing through the metallic conductor. In this situation the dipoles would be in disarray so that all magnetic interactions cancel out. Heat, that is nuclear vibrations, as well as the torques could be responsible for the dipole disorder. Ferromagnetic dipoles produce spontaneous order amongst themselves, however a piece of copper is non-magnetic and its dipole current elements are thus diamagnetic. A collection of 'permanent' diamagnetic dipoles has the property of inherent magnetic disorder. The significance of the elemental mutual torques becomes more evident in the next section which deals with electromagnetic induction.

Generalization of Neumann's Law of Induction

Neumann's elemental law of induction, Eq.1.39, applies to current elements of fixed orientation, constant current, and variable distance. For a more general validity of Neumann's theory it is necessary to establish formulae for the elemental induced e.m.f's when the angular orientation and the magnitude of the inducing current changes. To derive these formulae it will, first of all, be shown that any interaction between two Ampèrian current elements reduces to a two-dimensional problem. It is then sufficient to derive elemental induction formulae for co-planar element pairs.

It is always possible to put the locations M and N of two current elements in a common plane with the direction of one of the elements, say $i_m dm$ as shown in figure 1.3. Let x-y be the common plane and M the origin of a Cartesian coordinate system with $i_m dm$ pointing in the positive x-direction. If $i_n dn$, which makes the angle γ with the common plane, is resolved in parallel and perpendicular components $(i_n dn)_{x-y}$ and $(i_n dn)_z$ to this plane, the angles α, β and ε and Eq.4.4 apply to the co-planar elements $i_m dm$ and $(i_n dn)_{x-y}$. For the orthogonal pair, $i_m dm$ and $(i_n dn)_z$, $\cos\varepsilon$ and $\cos\beta$ are zero and therefore $\Delta P_{m,n}$ is zero regardless of any virtual displacement given to $i_n dn$. Thus no magnetic energy is stored between $i_m dm$ and $(i_n dn)_z$ and no mutual forces or torques exist between these orthogonal elements, whatever the orientation of $i_n dn$. This is in agreement with Ampère's rule stated between Eqs.1.5 and 1.6. We can thus ignore the orthogonal element combination since all current element interactions reduce to a two-dimensional model for which Eq.4.4 holds.

A pair of co-planar Ampèrian current elements have five degrees of freedom, which are the two current strengths, their distance of separation, and the inclinations of the two elements to the distance vector (see figure 4.2). It will now be shown that a time-variation of any of these five quantities generates an induced e.m.f. in the elements. The process requires four energy source-sink units. One is the store of magnetic energy which will be denoted by S. The elements communicate with their electrical energy supplies E_m and E_n. Finally, there must exist a mechanical work and energy agency which controls the distance of separation between the elements, which will be denoted by M.

(1)   *Variable $r_{m,n}$*

For attracting elements and increasing distance, M must supply energy to the elements. At the same time S loses energy. These two streams of energy have to flow into E_m and E_n which sustain the currents i_m and i_n. Joule heating plays no part in the conservative energy exchanges between M, S, E_m, and E_n shown in figure 4.5. Therefore heat will be left out of the analysis. If the induced electromotive force in a complete circuit is the sum of the elemental contributions, then each pair of current elements should provide

$$\Delta e_m = - \frac{d}{dt} \left(|\Delta M_{m,n}| \, i_n \right) \qquad (4.11)$$

where $\Delta M_{m,n}$ is defined as in Eq.3.19. In this instance, we are only interested in the magnitude of $\Delta M_{m,n}$, since it is that which varies inversely with r. The sign of the mutual inductance is simply a reflection of whether the force is attraction or repulsion. This will be called Neumann's generalized law of induction in terms of the mutual inductance between two Ampèrian current elements.

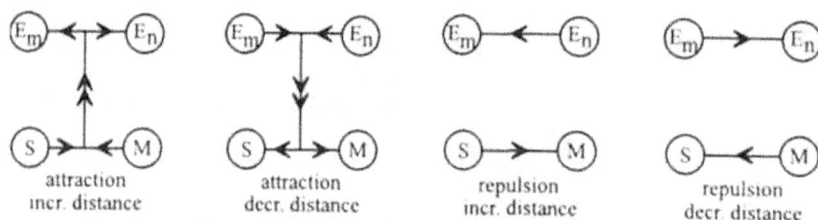

Figure 4.5 Power exchanges for varying distances

For increasing distance and constant currents and angles Eq.4.11 becomes

$$\Delta e_m = - \left(\frac{\partial}{\partial r_{m,n}} |\Delta M_{m,n}| \right) \left(\frac{d r_{m,n}}{dt} \right) i_n \qquad (4.12)$$

Let $v_r = dr_{m,n}/dt$ stand for the relative velocity of one element with respect to the other along the distance vector. With the Ampère interaction force $\Delta F_{m,n}$ we can write

$$\frac{\partial}{\partial r_{m,n}} \left| \Delta M_{m,n} \right| = \frac{\partial}{\partial r_{m,n}} \left[\left(\frac{dm\ dn}{r_{m,n}} \right) \left| (2\cos\varepsilon - 3\cos\alpha\cos\beta) \right| \right]$$

$$= - \left(\frac{dm\ dn}{r_{m,n}^2} \right) \left| (2\cos\varepsilon - 3\cos\alpha\cos\beta) \right|$$

$$= - \left| \frac{\Delta F_{m,n}}{i_m\ i_n} \right| \tag{4.13}$$

and the induced electromotive force becomes

$$\Delta e_m = \left| \frac{\Delta F_{m,n}}{i_m} \right| v_r \tag{4.14}$$

This agrees with Neumann's law of induction, as can be seen from Eqs.1.39 and 1.40.

The instantaneous power exchange between the mechanical source M and the electrical source E_m therefore is

$$\Delta e_m\ i_m = \left| \Delta F_{m,n} \right| v_r \tag{4.15}$$

Since the induced e.m.f. of Eq.4.15 is the result of relative motion it must cause the same energy exchange in both elements, such that

$$\Delta e_n\ i_n = \left| \Delta F_{m,n} \right| v_r \tag{4.16}$$

For this to be possible S must give up just as much power as M. Using Eq.4.3, the power flow from S is

$$\frac{d\left| \Delta P_{m,n} \right|}{dt} = \frac{\partial\left| \Delta P_{m,n} \right|}{\partial r_{m,n}} \frac{d r_{m,n}}{dt} = \left| \Delta F_{m,n} \right| v_r \tag{4.17}$$

which proves that energy is being conserved, and

$$\Delta e_m\ i_m + \Delta e_n\ i_n = 2\left| \Delta F_{m,n} \right| v_r \tag{4.18}$$

The same kind of analysis can be performed with a reversal of the relative velocity, that is for decreasing distance, and also for the two cases of repelling elements. The energy exchanges for the four different circumstances are summarized in figure 4.5. Let us briefly look at repelling elements and increasing distance. It requires S to supply energy to M. Calculations prove that all of the energy subtracted from the magnetic store is being absorbed by the mechanical source. There will still be e.m.f's induced in the elements in accordance with Eq.4.14, but they now facilitate an energy exchange between E_m and E_n. This is possible because one of the e.m.f's turns out to be a forward-e.m.f. and the other a back-e.m.f. The power exchanges depicted in figure 4.6 agree in every way with experimental observations on closed circuits.

(2)   *Variable α*

Instead of attracting and repelling forces, we now have to deal with the torque predicted by Eq.4.8. The e.m.f. equation applicable to variable α has to be derived from Eq.4.11. It is found to be

$$\Delta e_m = - i_n \frac{\partial \Delta M_{m,n}}{\partial \alpha} \frac{d\alpha}{dt} = - \frac{(\Delta T_{m,n})_\alpha}{i_m} \frac{d\alpha}{dt} \tag{4.19}$$

where $d\alpha/dt$ is the rate of change of alpha caused by the combination of the element $i_m dm$ rotating about its pivot and the element $i_n dn$ rotating about the element $i_m dm$. The electrical power flow in relation to the mechanical power is given by

$$\Delta e_m i_m = - (\Delta T_{m,n})_\alpha \frac{d\alpha}{dt} \tag{4.20}$$

The minus sign on the r.h.s. of Eq.4.20 confirms that the electromotive force must counteract the torque.

(3)   *Variable β*

The torque for this case is given by Eq.4.10. The e.m.f. equation applicable to variable β again has to be derived from Eq.4.11. It is found to be

$$\Delta e_m = - i_n \frac{\partial \Delta M_{m,n}}{\partial \beta} \frac{d\beta}{dt} = - \frac{(\Delta T_{m,n})_\beta}{i_m} \frac{d\beta}{dt} \tag{4.21}$$

where $d\beta/dt$ is the rate of change of beta caused by the combination of the element $i_n dn$ rotating about its pivot and the element $i_m dm$ rotating about the element $i_n dn$. The electrical power flow in relation to the mechanical power is given by

$$\Delta e_m \, i_m = - \, (\Delta T_{m,n})_\beta \, \frac{d\beta}{dt} \qquad (4.22)$$

(4) *Variable i*

Finally we examine the most common case of electromagnetic induction, that due to variable current intensity.

$$\Delta e_m = - \, \Delta M_{m,n} \, \frac{di_n}{dt} \qquad (4.23)$$

This formula involves i_n but not i_m. The other three e.m.f. equations -- Eqs.4.14, 4.19 and 4.21 -- contain both currents because they depend on mutual forces or torques which do not favor one element over the other. The value of the induced e.m.f. of Eq.4.23 does not depend on the strength nor the direction of any current flowing in the element that experiences the induction. In fact the latter current may be zero and the e.m.f. still exists. This is a major difference between statically and motionally induced e.m.f's. How can the sign of the mutual inductance be determined if the direction of one of the current elements is left undefined? For an answer to this question we have to fall back on experimental observations.

The line of action of the induced e.m.f. will normally be determined by the circuit (wire) layout, but it is unclear in which direction along this line the induced e.m.f. will tend to transport current. If we take two parallel wires and increase the current in one of them, then we know from experience that the induced e.m.f. in the other will point in the direction opposite to that in which the inducing current flows. This follows from Lenz's law and accounts for the negative sign in Eq.4.23. We may generalize the empirical finding and Lenz's law by saying that the direction of the induced e.m.f. in Eq.4.23 must be chosen such that $\Delta M_{m,n}$ is positive. No other rule appears to satisfy the law of induction.

The four induced e.m.f. equations -- Eqs.4.14, 4.19, 4.21 and 4.23 -- complete the generalization of Neumann's law of electromagnetic induction at the current element level. In this treatment the electromotive forces are given in e.m.u. of potential difference. The conversion factor from e.m.u. (cgs) to practical SI units is 1 volt = 10^8 e.m.u. As far as dimensions are concerned, it should be remembered that velocity is the e.m.u. of resistance. Therefore all the relative velocities in the e.m.f. equations could be converted to resistances by 1 Ω= 10^9 cm/s. It would, however, conceal the fact that induced electromotive forces are the result of relative velocities between current elements.

The mechanism of how the e.m.f produces charge transport remains a mystery. At the moment there seem to be two explanations. First, if the conduction electron is a part of the atomic dipole current element, then the swinging of this element about the atomic pivot could represent charge transport. Alternating currents may be partly or entirely due to swinging dipole elements. The second possibility is that the e.m.f. acting on an atomic dipole, once it has been aligned with the e.m.f., tends to increase the strength of the dipole. Forward e.m.f's would cause increases, and back-e.m.f's decreases, in this strength. For a sufficiently large e.m.f. the bond holding the conduction electron to the atom may break, allowing the electron to jump to the next ion. A back-e.m.f. may achieve the same result as a forward e.m.f., but with reversed current flow. Both e.m.f's could be assisted by applied (rather than induced) e.m.f's, supplied, for example, by batteries.

Torque Speculations

Since the discovery of electromagnetism by Oersted [1.2] in 1820 several new force concepts have been introduced to explain experimental observations. The first to appear was the ponderomotive Ampère force, closely followed by electromotive forces. The Lorentz force was later to embrace both the ponderomotive and electromotive force of electrodynamics. Since the time of Lorentz nearly a century has elapsed before more electrodynamic force concepts have come forward in the form of the alpha and beta torques. If they turn out to be real, why did it take so long to discover them?

Part of the answer is that generally taught electromagnetic theory conceals the existence of the two torques. It is conceded, however, that the practical consequences of the two torques are not at all obvious. In view of the many experimental failures of the Lorentz force there now exists every incentive to study the torques further. Solid progress in this field will depend critically on experimental tests and discoveries. In this book we can but speculate what the new forces may be able to do, and to what extent they may change our understanding of electromagnetism.

Recapitulating the current element story up to this point, we remember, first of all, that the Ampèrian current element had to involve the lattice ion in order to explain the existence of Ampère tension, and the tension fractures in wire explosions. Inductance calculations then made it necessary to associate each pair of atomic conductor elements with a finite amount of mutual inductance. In Neumann's theory of induction, this automatically assigned a certain quantity of stored magnetic (potential) energy to interacting current element pairs. With the principle of virtual work, the energy of current element pairs then confirmed Ampère's force law. At the same time this principle revealed the alpha and beta torques on current elements. Provided the atomic current element was a pivoted dipole and could swing freely about the lattice site, the torques turned out to be both electromotive and ponderomotive. This made it possible to account for all observed mechanical forces and induced e.m.f 's with the pivoted atomic current element model.

Experiment proves that electromotive forces can be induced in a neutral piece of metal, that is metal in which no current flows. In the generalization of Neumann's law of induction there arose circumstances when the induced e.m.f. was generated by torques acting on the atomic dipole elements. For this to be true, dipoles must exist even before the e.m.f. is induced. This argument leads to the conclusion that at all times, neutral pieces of metal must contain many permanent atomic current elements!

No unidirectional transport of charge occurs in the neutral metal, nor does it produce external magnetic effects. This behavior of the metal can only be explained if the permanent atomic dipole elements are in disarray and their magnetic effects cancel out. Chaotic charge transport could take place without the accumulation of charge in any region of the piece of metal. If we assume, however, in accordance with the quantum mechanical theory of electric conduction, that in normal metals charge transport always produces Joule heat, it appears that fluctuations in the random orientation of atomic dipoles does not result in the flow of conduction electrons from one atom to the next. Here it may be stressed once more that, in the Newtonian electrodynamics, magnetism is not the result of charge velocities but depends purely on the orientation of fixed current elements.

As well as the existence of the alpha and beta torques, thermal agitation of the metal lattice seems to be a likely cause of the disorder of the permanent dipole elements. A current would then be understood to be an alignment of dipole elements to form filaments along which the transport of conduction electrons can take place. Since the thermal agitation would continuously try to destroy this orderly alignment of dipoles, electromotive forces may have to exist all the time the current is flowing in order to maintain the dipole filaments. Thermal agitation would then require continuous work to be done on the dipoles. This work would have to be supplied from other energy sources "to drive" the current.

It is well known that the work done by a source of e.m.f. in driving current appears as Joule heat in the metal lattice. Turning a dipole element against the disordering thermal forces may generate heat which contributes to the thermal vibration of the nuclei at the lattice sites. It is far too early to inquire how this ohmic dissipation mechanism compares with the electron-lattice collisions underlying the quantum mechanical theory of electric resistance. The whole idea of permanent atomic dipole elements could flounder if it was found that the torque interactions predict spontaneous ordering of the elements, resulting in collective magnetism for which there is no experimental evidence. The issue of spontaneous ordering will be considered next.

Filaments of aligned atomic elements, not transporting charge, could be called a diamagnetic current. Qualitatively, this current would seem to produce the same magnetic effects we attribute to ordinary electric current. Neugebauer [4.5] and others called this absolute diamagnetism to distinguish it from the conventional induced diamagnetism. It will be recalled that induced diamagnetism is responsible for a weak magnetic field which opposes the inducing field. In superconductors, however, this effect becomes very prominent and leads to the cancellation (magnetic flux expulsion) of the applied magnetic field in the body of the superconductor by the flow of 'persistent' currents in the skin of the body. This is the Meissner effect. Neugebauer speculated that it may be due to aligned ion chains. They could be current filaments of the Ampère-Neumann electrodynamics.

There exists no experimental proof that persistent currents in superconductors actually involve charge transport. In order to prove this, a metallic superconducting circuit would have to be interrupted. This would stop the motion of the electrons, but at the same time would probably destroy the magnetic order that makes up a filament, particularly if this order depends on near-neighbor interactions. Therefore, opening a persistent current filament is likely to stop charge transport as well as the diamagnetic current.

The thermal disordering would presumably be absent in a superconductor. Hence if it is possible to establish a certain dipole order in superconductors, it would probably persist as long as no external magnetic disturbance occurs. To investigate the ordering effect we assume that no current flows in the conductor. A pair of current elements, $i_m dm$ and $i_n dn$, would then tend to find angular positions at which the collective actions of the alpha and beta torques are zero and the elements are stable to small perturbations. Figure 4.4 shows that this is true for a pair of elements when $\alpha = -\beta = (90°$ or $-90°)$.

The situation is more complicated however when considering the mutual torques exerted by a large number of elements. Calculations have been performed on a 20×20 planar array of elements, revealing that there are stable positions for these elements as a result of the torques predicted by Eqs.4.8 and 4.10. The elements tend to form radial current filaments emanating from the centre of the conductor. It appears that all of the elements pointing inward as well as all of the elements pointing outward are both stable arrangements. The latter case is depicted in figure 4.6. Since there is no source or sink of charge at the centre of the superconductor, these current filaments do not represent charge transport. However it appears that their existence is consistent with the concept of no magnetic field in the body of the superconductor as predicted by the Meissner effect.

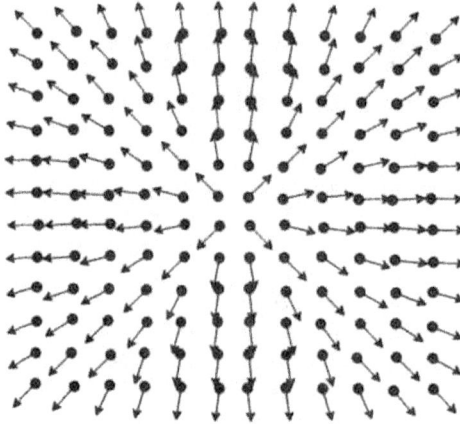

Figure 4.6 : Stable element directions for an element array

External currents will still be able to affect the diamagnetic dipole elements in the skin, and in connection with this it should be noted that the London theory of superconductivity [4.6] indicates that inside a singly connected superconductor the magnetic flux density B obeys the diffusion equation

$$\nabla^2 B = \lambda^2 B \qquad (4.24)$$

where λ is a material constant determining the depth in which the persistent surface currents flow. These currents are a feature of the Meissner effect [4.6]. Equation 4.24 actually requires that the magnetic field is expelled from the interior of the superconductor and persists in a thin surface layer. The expulsion of the magnetic flux is taken to be the physical manifestation of the Meissner effect.

From figure 4.4, it can be seen that as a result of the predicted torques, the elements tend to counteralign themselves. This makes it plausible that these torques are the origin of the counteraligned persistent skin currents in a superconductor created by an external current loop. As one moves into the body of the superconductor, the effect of the external current becomes weaker due to increasing distance from the external current, and screening from the outer elements. The level at which there is negligible effect from the external current defines the skin depth, below which the elements would point radially as in figure 4.6.

Since the number of atomic current elements in the superconducting body is finite, there must come a time, with increasing external current density, when all elements are aligned with the skin currents, representing diamagnetic saturation and thereby eliminating the Meissner effect. The pivoted dipole model of the current element, together with the alpha and beta torques, therefore provide a new explanation for the quenching of superconductivity by what would otherwise be called the critical magnetic field strength.

Any kind of similar diamagnetic order in normal metals is expected to be continuously upset by thermal agitation. This agitation must be capable of turning pivoted dipole elements. Hence normal metals should experience a very weak Meissner effect. This could be the phenomenon which is ordinarily called 'diamagnetism'. The argument is consistent with the experimental fact that ordinary superconductivity occurs only at low temperatures.

There is evidence that superconductivity can also be quenched by applying tension to the conductor. Remembering that in Chapter 2 there was experimental and theoretical understanding of how tension is created in any current carrying conductor as a result of longitudinal Ampère forces, it seems natural therefore that current in a superconductor may create internal tension. If this current were to exceed a certain value at which the internally generated tension exceeded its critical value, then superconductivity would be quenched. This could well be the source of the critical current limit, which in the case of Type II superconductors appears to operate independently of the critical field strength limit.

If it is assumed that the electrical resistance of metal is caused by thermal agitation and turning of pivoted current elements, it is quite remarkable how virtually all of the aspects of superconductivity can be explained with the Newtonian electrodynamics. It can account qualitatively for a critical transition temperature, the appearance of zero resistivity, persistent current skin formation, and quenching by a critical current or field.

In the London theory, the magnetic vector potential obeys an equation which is similar to Eq.4.24. This immediately suggests that Neumann's theory, and therefore the Newtonian electrodynamics, can also explain the Meissner effect, albeit without invoking a magnetic field. Both superconductivity models -- those of relativistic electromagnetism and the Newtonian electrodynamics -- cannot be right. Which is correct may be resolved with an easy laboratory test, which is analogous to the Aharonov-Bohm demonstration [4.7, 4.8] of action at a distance in quantum mechanics. The superconductivity test was first suggested by Peter Graneau in 1985 [1.12], however ten years later it has still not been performed.

Consider an ideal (type I) superconducting ring of, say, lead. When this is cooled through the superconducting transition temperature, the earth's magnetic field should be expelled from the lead. If a coaxial turn of copper wire is placed near the ring, and twisted leads from the wire are brought out of the cryostat to an oscilloscope, a faint voltage pulse may, or may not, be observed on the oscilloscope screen, indicating the field expulsion from the lead and the switch-on of a weak persistent current. To make the induced voltage pulse strong and unmistakable, the superconducting ring should be placed in the field of an electromagnet.

For the Aharonov-Bohm experiment, a long solenoid carrying a substantial DC current should be used. The lead ring should encircle the solenoid in its center plane. Except for a thin layer of insulation, the superconducting ring must touch the solenoid. Under these conditions, as Aharonov and Bohm outlined [4.7], the magnetic field at the location of the lead ring is very small but the magnetic vector potential is very large. If a copper wire turn close to the lead ring detects a significant induced voltage at the temperature transition, it would prove that the magnetic vector potential is the cause of the Meissner effect, confirming the prediction of the Newtonian electrodynamics of superconductivity.

If no significant voltage pulse is induced at the transition temperature, the experiment would be less decisive. It could then be that the superconducting transition was very slow providing an insufficient rate of change of the vector potential for an induced voltage pulse to be detected.

The magnetic vector potential arose first in Neumann's theory of induction and, as Eq.1.47 proves, it is closely associated with Ampèrian current elements. Consequently, strong interactions between the dipole elements of the solenoid and those of the superconducting ring should make themselves felt.

The experiment should be performed with constant DC current in the solenoid. The assembly of solenoid, superconducting ring, and copper wire loop should then be cooled down with liquid helium, while a steady current flows through the solenoid. Provided the temperature falls quite rapidly, the switch-on of the persistent current should induce a strong e.m.f. pulse in the copper wire loop.

In reference [1.12], the analysis of a superconducting filament surrounding a normal current filament was extended to the case of the superconducting filament surrounding a solenoid. For a mutual inductance M between the solenoid and the single filament turn of radius R, it was found that the persistent current i_m relative to the solenoid current i_n was given by

$$\frac{i_m}{i_n} = -\frac{M}{2\pi R} \qquad (4.25)$$

There are techniques available for measuring and computing M. Equation 4.25 may possibly be tested by experiment.

A Curious Coincidence

Nothing has been said in this book about the interaction of Ampèrian current elements with magnetic dipoles. These interactions certainly exist and must lead to the largely unexplored world of Newtonian magnetism. It is a fascinating subject as the following exercise will demonstrate.

Two permanent Ampèrian current elements can be made out of magnets. By hand-holding the two elements one can feel all the ponderomotive interaction forces of the Newtonian electrodynamics. All that is needed are two permanent rod magnets, as shown in figure 4.7(a). Draw an arrow on one of the magnets from the south pole S to the north pole N. On the other magnet reverse the arrow so that it points from N to S.

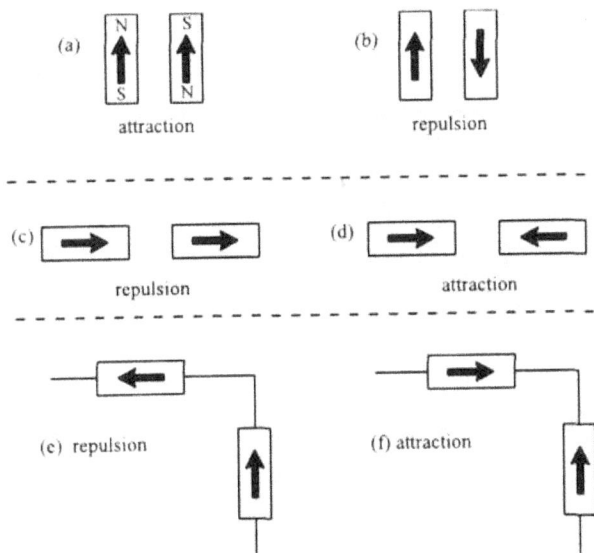

Figure 4.7 : Artificial Ampèrian current elements

Now concentrate only on the arrows, for they take the place of two Ampèrian current elements. When they are laid side-by-side, as in figure 4.7(a), they attract each other just like real current elements. When one is reversed, as in figure 4.7(b), the elements repel each other, again like two current elements would repel. In the in-line configurations of figure 4.7(c) and (d) the magnets again repel and attract like Ampèrian current elements. In figure 4.7(e) and (f) the elements are arranged across the corner of a circuit. The railgun driving force is demonstrated by (e), and (f) reveals the attraction of currents converging on a point. Of course the equations are still needed to determine the magnitudes of all forces and torques. Nevertheless, the artificial current elements are good qualitative guides through the mathematical maze of the Newtonian electrodynamics.

Is this behavior of two permanent rod magnets a coincidence which has nothing to do with the laws of Ampère and Neumann? Probably not. There is no doubt that the static Ampèrian current element has magnetic properties. In fact that is all it is: some kind of unusual magnet. On a few occasions the elements have been described as permanent diamagnetic dipoles. A microscopic definition of this kind of dipole is still lacking. Worse than that, we are at a loss to describe the phenomenological difference between diamagnetic and paramagnetic dipoles. But as the experiment described in this section proves, paramagnets can be used to mimic diamagnetism.

Chapter 4 References

4.1   J.P. Wesley, *Causal quantum mechanics*, Benjamin Wesley, Blumberg (Germany), 1983.

4.2   H.A. Lorentz, *The theory of electrons*, Teubner, Leipzig, 1909.

4.3   A. Einstein, "On the electrodynamics of moving bodies", The principle of relativity, Dover, New York, 1923.

4.4   F.F. Cleveland, "Magnetic forces in a rectangular circuit", Philosophical Magazine, Vol.21, p.416, 1936.

4.5   T. Neugebauer, "Zu dem Problem des absoluten Diamagnetismus und der Supraleitung", Acta Physica Hungaria, Vol.17, p.203, 1964.

4.6   F. London, *Superfluids*, Vol.1, Dover, New York, 1961.

4.7   Y. Aharonov, D. Bohm, "Significance of the electromagnetic potentials in quantum theory", Physical Review, Vol.115, p.485, 1959.

4.8   G. Moellenstedt, W. Bayh, "Messung der kontinuierlichen Phasenschiebung im kraftfreiem Raum durch das magnetische Vektor Potential", Naturwissenschaften, Vol.49, p.81, 1962.

CHAPTER 5

The Railgun : Testbed of the Newtonian Electrodynamics

Description of Railguns

Birkland [5.1] appears to have been the first to construct an electromagnetic gun. He used switched solenoid sections to accelerate ferromagnetic projectiles. In 1903 he arranged a demonstration which was supposed to be silent but instead produced a loud arc explosion. Today this first electromagnetic launcher would be called a coil-gun.

Railguns are almost as old a concept as coil-guns. They accelerate metallic conductors which bridge a pair of current rails. Sliding metallic bridges and electric arcs in air have been used as the driven armatures which transfer current from one rail to the other. Both alternating and direct currents will drive railguns, but DC is less wasteful in Joule heating. Current pulses are used for the most powerful shots. The first major railgun experiment was performed in Australia in the 1970s [5.2]. Since then the railgun has become a potential weapon as part of the US Strategic Defense Initiative (SDI) as well as a space launcher for placing objects in orbit around the earth.

Figure 5.1 serves to further explain railgun terminology. DE and CF are the rails. DC is the breech branch which contains the power source. The latter may simply be a battery, however powerful railguns are driven by capacitor banks or rotating generators. The armature AB is accelerated by the electrodynamic force F_L and travels with increasing displacement x and velocity v from the gun breech DC to the muzzle EF, and then leaves the gun.

V_b is the breech voltage and V_m the muzzle voltage. In the conversion from electrical to mechanical energy, which takes place in the railgun because the electrodynamic force F_L does work, there must arise an induced back-e.m.f. in the circuit. Thus, as well as Joule heat, the electric power source must also provide energy by driving the current i in opposition to the back-e.m.f., e_b.

In practice it has been found that very sturdy rails are required. They are usually placed close together so that the armature is more bullet shaped. Figure 5.1 is merely a circuit diagram and not a drawing of a railgun.

In the design and operation of railguns there arise three interesting questions which cannot be adequately answered with field theory. They are:

(1)   How is the driving force F_L produced?

(2)   Where in the circuit does the recoil force arise?
(3)   What is the distribution of the induced back-e.m.f.?

The railgun is probably the least effective of all electromagnetic launchers, and is certainly outperformed by the water-arc gun to be described later. The purpose of analyzing railguns here is not to examine their detail and utility, but to illustrate that these important electromagnetic machines cannot be understood without the Newtonian theory outlined in this book.

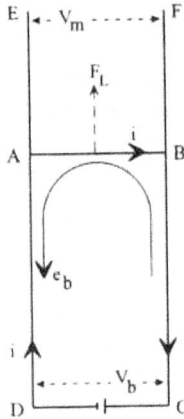

Figure 5.1 : Single-filament railgun model

Nonlocal Action

The many textbooks dealing with relativistic electromagnetism are remarkably silent on the issue of how the electrodynamic forces on metallic current-carrying conductors are produced. The magnetic pressure cannot arise from mere static contact with the magnetic field, for this would truly represent action at a distance. While Newtonian universal gravitation may be associated with a static force field, this type of mechanism does not produce the Einstein local action required by relativistic electromagnetism. The difference between a force field and the electromagnetic field of the Maxwell-Einstein theory is the presence of field energy which must travel at the velocity of light. In electrodynamic situations the field energy is always moving with speed c (the velocity of light), even when constant DC current is flowing.

This theory leads to energy motions which defy common sense. For example, when a constant electric current flows along a straight wire, field theory requires that energy from the current source flies through the air, or vacuum, and enters each section of the wire from the outside. The radially incoming energy is responsible for setting up the e.m.f. which drives the current against the metal resistance. On this point Feynman [5.3] observed:

"So our "crazy" theory says that the electrons are getting their energy
to generate heat because of the energy flowing into the wire from the field
outside."

The special theory of relativity claims that travelling field energy possesses momentum
and mass called 'electromagnetic mass'. As the field energy strikes metal and penetrates into
its depth, it is arrested and loses its momentum. Given that momentum must be conserved
whatever theory is being used, this requires the transfer of momentum from mass 1 to mass 2
which can only be achieved provided masses 1 and 2 are subject to equal and opposite
reaction forces. In this way relativistic electromagnetism comes to predict the Lorentz force,
F_L, which accelerates the railgun armature, while the local reaction force on the incoming
electromagnetic mass brings the energy velocity down to the velocity of the metal. The
reaction force to the Lorentz force, which stops the incoming energy, thus takes on the role
of the field recoil force (magnetic pressure) of the railgun shot.

As Feynman [5.3] explains, the Poynting vector ($\vec{S} = \vec{E} \times \vec{H}$) of the travelling
energy divided by the square of the velocity of light c, represents the momentum flow density
$\vec{g}$, given by

$$\vec{g} = \frac{1}{c^2} \vec{S} \tag{5.1}$$

Arresting energy-momentum flow with a metal barrier gives rise to the magnetic component
of the Lorentz force and, thereby, generates magnetic field pressure.

This is the local action mechanism on which Einstein insisted in order to do away with
what he called 'spooky action at a distance'. He also provided his famous energy law which
applies to the generation of magnetic pressure. If m_e is the electromagnetic mass which
travels with the velocity of light, then the energy E associated with this mass is

$$E = m_e c^2 \tag{5.2}$$

If m is the mass of the railgun projectile which leaves the muzzle with velocity v, then
momentum conservation predicts that

$$m v = m_e c \tag{5.3}$$

The implication of this local force mechanism is best understood by considering a
numerical example. In 1984 Deis, Scherbarth and Ferrentino [5.4] reported an actual railgun
shot in which 16.3 MJ of stored electrical energy accelerated a 0.317 kg projectile to a
velocity of 4200 m/s. Under the action of the local Lorentz force, therefore, the projectile
acquired a momentum of 1331 kg.m/s. Using Eq.5.3, momentum conservation predicts that
the projectile had been struck by at least 4.44×10^{-6} kg of electromagnetic mass travelling at
the velocity of light ($c = 2.98 \times 10^8$ m/s). Then on the basis of Einstein's energy law, Eq.5.2, this
demanded that the total energy which travelled between the rails from the energy source to

the armature was 3.99×10^5 MJ, or 24,478 times the energy actually delivered to the railgun.

From this example it must be concluded that the magnetic force on the railgun armature cannot be produced by field-energy impact. Here we have a practical example which reveals a serious flaw of relativistic electromagnetism.

The kinetic energy which the projectile of mass m, and muzzle velocity v, acquired is, of course,

$$E_k = \frac{1}{2} m v^2 \qquad (5.4)$$

It comes to 0.559 MJ. Hence the energy efficiency of the EMACK railgun of Deis et al was less than four percent. Much of the wasted energy probably contributed to Joule heating of the circuit and friction losses between armature and the rails [5.5].

If the energy absorbed by the projectile from the field had been equal to the kinetic energy acquired, then from Eq.5.2 the electromagnetic mass which should have struck the projectile (armature) would have been 6.21×10^{-12} kg. Multiplying this by the velocity of light gives the field-momentum of 1.86×10^{-3} kg.m/s. This is far smaller than the 1331 kg.m/s momentum of the projectile. Hence if energy is conserved, momentum is not, and vice versa. This is the greatest inconsistency of relativistic field theory.

The same problem arises in relativistic quantum mechanics, also referred to as quantum field theory or quantum electrodynamics. In order to explain the simple Newtonian repulsion of two electrons in vacuum, quantum electrodynamics (QED) teaches that the electrons exchange virtual photons. Each electron must spontaneously and continuously emit a stream of photons. As these particles of light contain electromagnetic mass, they exert a recoil force on the emitting electron and an impact force on the absorbing electron. These are the locally generated forces of special relativity. As no external agency causes the energy emission, the photons clearly violate energy conservation. This is considered to be permissible so long as the life of the virtual photon is so short that its energy is subject to the uncertainty principle of quantum mechanics.

Allen and Jones [5.6] tried to rescue the local action principle for railguns. They saw no difficulty in using electromagnetic mass and momentum to calculate the projectile acceleration force. To do this they assumed that the field energy was multiply reflected between the armature AB and the breech DC of figure 5.1. Even though the net energy flow density (the Poynting vector) near the reflector was zero, these authors maintained that after just one total reflection, the magnetic pressure would still be exerted on the reflector for the same reason that a gas exerts mechanical pressure on the container walls, while the total momentum in any given gas volume element is zero. We note that no part of such a gas container moves and in this Newtonian situation momentum is conserved. The analogy, however, does not apply to railguns where the reflector is being accelerated and therefore requires the transfer of momentum.

Before reflection, a certain electromagnetic mass m_e would have a forward momentum $m_e c$. After reflection this mass has an oppositely directed momentum $-m_e c$. Assuming that the reflecting body, that is the armature, of real mass M, has been accelerated by the reflection to the forward velocity v, then the momentum balance before and after the reflection is

$$m_e c = M v - m_e c \tag{5.5}$$

or

$$M v = 2 m_e c \tag{5.6}$$

In other words, the mass M was given twice the momentum it would have received had the electromagnetic energy $m_e c^2$ been absorbed instead of reflected. If this were true, arbitrarily large accelerations forces could be generated by repeated to-and-fro reflections. A small amount of energy could thus produce large momentum. Once more the Newtonian simultaneous conservation of energy and momentum would be sacrificed.

The Allen and Jones proposal also ignores Poynting's theorem which, in field theory, governs the flow of electric currents in metallic conductors. With this theorem field theory actually demands the total absorption of incident field energy in order to supply the Joule heat dissipated in the conductor. Poynting [5.7] wrote:

"It seems that none of the energy of a current travels along the wire, but that it comes from the non-conducting medium surrounding the wire, that as soon as it enters it begins to be transformed to heat, the amount crossing successive layers of the wire decreasing until by the time the center is reached, where there is no magnetic force, and therefore no energy passing, it has all been transformed into heat. A conduction current then may be said to consist of the inward flow of energy with its accompanying magnetic and electromotive forces, and the transformation of energy into heat within the conductor."

Allen and Jones simply ignored the Joule heat altogether and wrote [5.6]:

"We note that the fraction of incident power that is not reflected from the system is partly 'stored' and partly transformed into mechanical energy."

Primarily because of Poynting's theorem, modern electromagnetism does not allow the magnetic component of the Lorentz force to do **any** mechanical work, such as accelerating the railgun armature, for all of the incident energy is immediately converted to heat. The same conclusion was reached in a study of induction motors [5.8].

In their justification of the energy reflection mechanism, Allen and Jones treated the railgun as a transmission line. This appears to be perfectly permissible so long as there are energy waves or voltage disturbances transmitted along the line. Wave equations are the basis of all transmission line theory. The wave equations vanish, however, when the current is constant, but railguns are known to operate with steady DC currents. This fact alone is sufficient to prove that transmission line theory is inadequate to explain the operation of railguns.

Finally, if a block of energy travels many times to-and-fro along a transmission line, it has to dissipate some of its content in the rail resistance (attenuation). The total number of reflections which should take place and determine the acceleration force should, therefore, depend on the rail resistance. On account of this, and for the same constant current, copper rails should produce more force than aluminum rails. No experiment has ever shown this effect.

Allen and Jones [5.6] went on to consider quantized energy flux along the railgun. This does not eliminate the Poynting theorem, nor does it alter their reflection mechanism. They did not, however, mention that quantum mechanics has revealed the existence of nonlocal actions. This was first discussed by Aharonov and Bohm [4.7], and experimentally confirmed by Moellenstedt and Bayh [4.8]. Bell [5.9] arrived at the same conclusion by logic and published the Bell inequalities. Bell's nonlocal actions, implied by distant quantum correlations, were experimentally confirmed by Aspect et al [5.10]. It now appears that railguns are an example from classical physics which, like quantum processes, require nonlocal actions to describe their behavior. The local Lorentz force theory fails because it does not conserve energy.

Vigier [5.11] suggested that the energy reflection hypothesis of Allen and Jones could be tested experimentally by enclosing the armature in a metal box. The experiment was performed by Neal Graneau at Oxford University. He used two half-inch diameter copper pipes as rails and laid a quarter-inch diameter stainless steel rod across the rails. The latter would act as armature and roll away from the breech. A 12 V car battery served as the current source. As shown in figure 5.2, the rails were passed through an aluminum metal box. This consisted of a cardboard box covered with aluminum foil. The foil was connected to the rail conductors, leaving no gaps, and therefore carried some current. The box had a removable lid so that the armature could be placed on the rails inside the box.

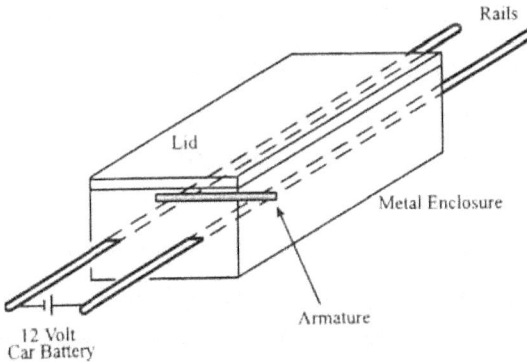

Figure 5.2 : Metal enclosure surrounding railgun armature

The Poynting energy flow from the battery should, according to Allen and Jones, be reflected by the aluminum and never reach the armature. Hence no electrodynamic force should have been exerted on the armature and it should have remained stationary in the box when the battery was connected to the rails. Contrary to this hypothesis, the rod did roll inside the aluminum box, thereby disproving energy reflection, not only by theory, but also by experiment. At the same time the experiment confirmed the action at a distance forces of Ampère's law which cannot be suppressed by shielding.

Recoil Force

If it had been possible to substantiate local Lorentz forces on the railgun armature, the recoil would have been felt by the field energy approaching the armature. In addition to this action and reaction at the armature, other local forces on the battery, or power supply, should have produced a kick-back as the driving energy was emitted into the gap between the rails. Hence local action predicts two recoil forces, one associated with emission and the other with absorption of field energy. It is important to realize that the force on the current source in the breech is not the recoil to the energy absorption force on the armature. In the previous section it has been shown that the local force mechanism cannot possibly exist, and therefore the recoil question has to be examined anew.

The difficulties which conventional electromagnetic theory has with the recoil forces are immediately apparent from the scarcity of publications dealing with this subject. Apart from the authors' railgun recoil papers [5.5, 5.12, 5.13], only two other treatments of this subject have been found amongst the hundreds of railgun publications of the past ten years. This is strange because the recoil force must be as large as the acceleration force, and the latter can amount to 100,000 N. The other recoil papers were written by Weldon et al [5.14] and Robson and Sethian [2.26].

Weldon et al claimed that the railgun recoil force was the Lorentz force on the breech conductor, and not just on the power source, or the battery in figure 5.1. As Poynting's theorem does not allow for the emission of field energy from the breech conductor, it is clear that Weldon et al, perhaps unwittingly, were considering nonlocal (that is action at a distance) forces. Had they assumed local action, the force should have been experienced only by the current source and not the rest of the breech conductor.

To resolve the recoil problem it is necessary to examine the reaction force pair, and particularly the reaction force distribution, between the two parts of the railgun circuit, that is (1) the armature AB and (2) the rigidly interconnected rail-breech combination ACDB of figure 5.3. Differences show up clearly when one looks at the longitudinal reaction force distribution given by the Lorentz and Ampère force laws. Both would have to depend on action at a distance because the breech conductor is unable to emit field energy.

Figure 5.3(a) shows the Lorentz reaction force distribution on which Weldon et al relied. The total Lorentz force on the armature, F_{AB}, is equal to the total Lorentz breech force, F_{CD}. They comply with Newton's third law for the circuit as a whole, provided they are simultaneous far-action forces. However the Lorentz interaction between one current element in the rail AC and one element in the breech CD is not simply attraction or repulsion, which can lead to conflict with Newton's third law. In fact almost all of F_{CD} is generated by interactions with current elements in the rails AC and BD. Hence F_{CD} is really a self-force arising from internal interactions between current elements of the same rigid conductor. This self-force mechanism, described in detail in Chapter 2, is inconsistent with Newton's third law. For this reason alone it is unlikely to be correct. Of the many publications which deal with the self-force dilemma of field theory, Cleveland's [3.3] is particularly lucid.

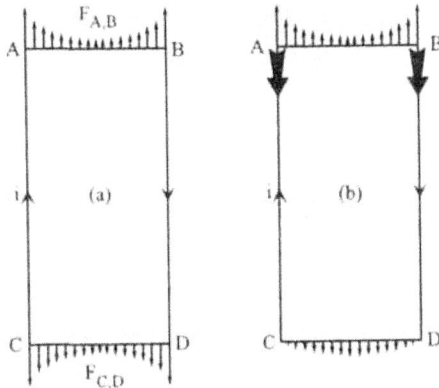

Figure 5.3 Longitudinal reaction forces between ACDB and the armature AB according to (a) the Lorentz force law, and (b) Ampère's force law

Since the recoil mechanism suggested by Weldon et al is clearly a far-action mechanism, it should result from a far-action force law such as Ampère's law. However the latter predicts that the recoil force has its seat in the rails, and most of it should be located quite close to the armature, as shown in figure 5.3(b). The Ampère recoil arises from the current element repulsion across the circuit corners at A and B. This repulsion is shown in detail in figure 5.4. The current element m in the armature branch P repels the current element n of rail S by the force $\Delta F_{m,n}$ of Eq.1.24, Ampère's force law. This force is positive (repulsion) because, in this particular case, cos $\varepsilon = 0$ and cos α and cos β are positive. The transverse component of the repulsion of m is a contribution, ΔF_a, to the armature acceleration force, while the longitudinal component of the repulsion force on n, ΔF_r, is a contribution to the rail recoil force. This holds for any current element combination between armature and rail. A similar Ampère recoil force will be developed in the other rail. It is difficult to depict the rail recoil force distribution diagrammatically. In figure 5.3(b) this has been attempted with bold arrows near A and B.

To demonstrate just how concentrated the recoil forces are in the two rails, their distribution has been calculated. For the purpose of the example, the armature AB (see figure 5.1) was 30 cm long and the rail length was AD=BC=100 cm. These dimensions were chosen to be equal to those of the experimental circuit of Chapter 3, to which figures 3.1 and 3.3 refer, and on which force measurements were carried out.

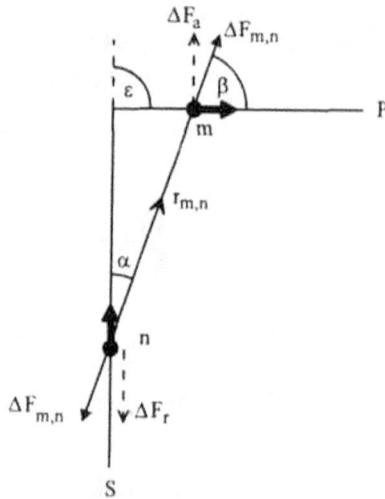

Figure 5.4 : Ampère recoil force ΔF_r between a current element in the armature branch P and another element in the rail S

The calculation employed the rectangular strip shown in figure 3.4(b) and (c), of 0.5 inch width and 0.05 inch thickness. As shown in this figure, the strip was divided into ten parallel filaments. Each filament was further divided into cubic current elements. One 100 cm long rail then consisted of 7870 elements.

The recoil force was calculated for different values of m, where m is the element position along one of the rail edge filaments, starting at the armature end. As shown in figure 3.4(c), there are ten elements arranged side-by-side which all contribute to the specific recoil force $\Delta F_{r,m}/i^2$ at the particular m-value, where i is the total strip current. The results are presented in figure 5.5.

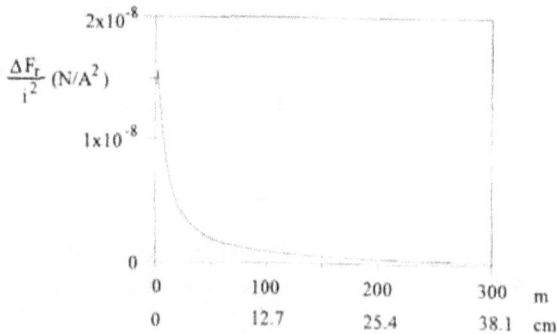

Figure 5.5 : Ampère recoil force distribution along the first 38 cm. of a 100 cm. long rail

It was found that a little over 71 percent of the recoil force in one strip rail has its seat in the first ten centimeters behind the armature. This distribution implies that the rail will be susceptible to buckling. The recoil force distribution is, of course, the same in both rails, and the sum of the two rail recoil forces is almost the same as the acceleration force on the armature. This leaves less than one percent of the recoil to be felt by the breech of the rectangular circuit.

Assuming a current pulse of 100 kA amplitude, the 71 percent recoil on the first ten cm amounts to 3330 N. To what extent this force will buckle the strip must depend on the duration of the current pulse. Inertia of the strip atoms prevents instantaneous deformation. Strip buckling under railgun recoil has been confirmed by experiment.

Before describing this test, the transverse force distribution on the armature will be examined. This is of interest because it represents an instance where there is excellent agreement between the Lorentz and Ampère formulae. Furthermore, the total transverse force on the armature must be equal to the total recoil force on the remainder of the circuit. In this way transverse force calculations afford a check on the recoil force calculations.

The transverse forces on the 10-filament strip armature peak at the two ends. Table 5.1 lists the results of transverse force calculations for the Ampère and Lorentz force formulae. The forces are tabulated in (dyne/ab-amp^2). As mentioned previously, these forces can be converted to (N/A^2) by multiplying by 10^{-7}. The Lorentz force is given by Grassmann's law, Eq. 1.58. Since the force distribution is symmetrical about the center line of the railgun, only half of the element positions have been listed in table 5.1. Any specific m-value stands for the position of the element in an edge-filament. The listed force acts on the combination of this and the nine adjacent elements.

m	Ampère Force (dyne/ab-amp^2)	Lorentz Force (dyne/ab-amp^2)
1	0.37538	0.26962
2	0.23261	0.23242
3	0.19884	0.19913
4	0.17360	0.17389
5	0.15363	0.15393
6	0.13744	0.13774
7	0.12409	0.12438
8	0.11293	0.11322
9	0.10349	0.10379
10	0.09544	0.09574
40	0.02967	0.03000
118	0.01637	0.01674
Total	4.70148	4.6342

Table 5.1 : Calculations of transverse force distribution on the armature

It will be noted that the two distributions are very similar and the total transverse forces differ by less than 1.5 percent. The measure of agreement between the Ampère and Lorentz forces in this example is quite remarkable. A distinguishing feature is the contribution of the end elements. It is possible that in reality the two distributions are identical and the small differences in table 5.1 are due to errors produced by using macroscopic rather than atomic current elements.

From table 5.1 half the total specific acceleration force on the armature comes to 4.70. The total force therefore is 9.40 which has been plotted on the graph of figure 3.3 and is seen to be very close to the measured force on the armature. The calculations that produced the results shown in figure 5.5 show that the total rail recoil force is 9.38, leaving no more than 0.02 for the recoil action on the breech.

Recoil rail buckling was demonstrated with the simple experiment of figure 5.6. The rails were supported on the outside by wooden beams, D, so that the transverse force on the rails could not deflect them outward. The main portion of the rails, A, consisted of 0.5-inch-wide, 0.05-inch-thick copper strip secured to wooden beams up to 30 cm behind the stationary armature, a. The last 40 cm of the rails, B, consisted of much thinner strips of the same width as the thick rails. Both aluminum and stainless steel were used for the thin rail extensions. The latter were pinned at p to the thick copper rails and the beams. A 0.5-inch-diameter copper rod fixed in position, formed the armature, a, and was in light contact with the thin rails.

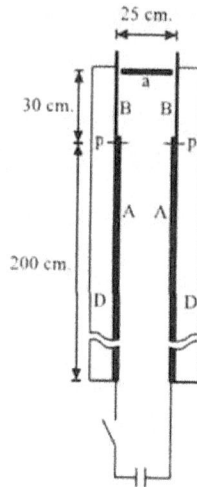

Figure 5.6 : Rail recoil buckling experiment

An 8-μF capacitor bank, charged to various voltages up to 80 kV, was discharged through the railgun setup in which the rails were spaced 25 cm apart. Current pulse amplitudes varied up to 100 kA. With sufficient current to heat the thin rail portions to within a few hundred degrees of their melting points, the strips, B, were found to deform plastically in two buckling modes. They retained their distorted shapes during cool-down for subsequent inspection and photography. The simple inward deflection of figure 5.7(a) was obtained with aluminum rails. Stainless steel rails buckled in concertina fashion, as can be seen from figure 5.7(c). Only the existence of longitudinal Ampère forces can adequately explain the observed rail buckling.

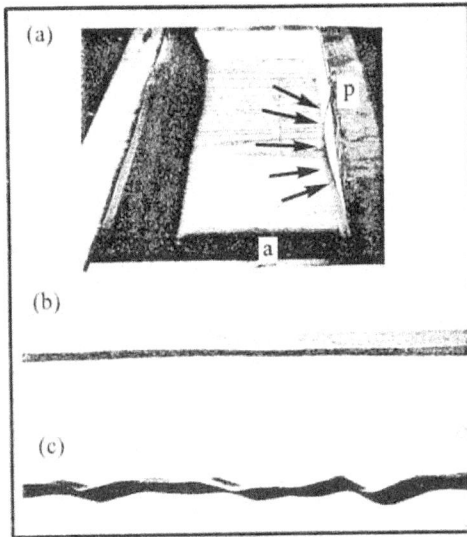

Figure 5.7 Buckling of thin rails: (a) inward deflection of aluminium rail; stainless steel rail (b) before and (c) after recoil experiment

Robson and Sethian [2.26] claimed that the strip distortions shown in figure 5.7 took place when the strip was softened by raising its temperature to red heat with the current pulse. They suggested that residual stresses, retained from cold-working and coiling of the strip, were unable to deform the cold strip (figure 5.7(b)), but would do so as soon as the strip was heated. The distorted shape would be retained during cooling. The Robson and Sethian argument assumes that red heat will not anneal (recrystallize) the strip, for after annealing there is no residual stress left that could be relieved. The Robson and Sethian mechanism of strip deformation contradicts metallurgical knowledge. Recrystallization cannot be prevented and the mere heating of a piece of metal increases the distance between atoms without deformation of the metal structure. In any case, it can easily be shown that if a straight cold piece of stainless steel strip, like the one shown in figure 5.7(b), is laid on a tray in an annealing furnace and then heated below or above the annealing temperature, it will not deform to the shape shown in figure 5.7(c).

The experiment of figure 5.6 provided a second proof of the existence of longitudinal recoil forces in the rails. When the thin rail extensions were not perfectly aligned with the copper rails, the ends near the armature would be pushed up or down, pivoting about the pins, p. This up or down swinging of the rail sections is certainly not a thermal effect, but it is inevitable if recoil forces push the rails back. The enormously strong recoil forces of military railguns will buckle and deflect the rails in both the elastic and plastic mode. This is likely to cause interference between armature and rails which gives the impression of severe friction, and probably explains the low mechanical efficiency of railguns [5.5].

Robson and Sethian [2.26] also performed an experiment which they thought disproved the existence of Ampère recoil forces in railguns. It therefore becomes necessary to show the flaws in their reasoning. This experiment has been briefly discussed in Chapter 2.

A simplified version of the Robson and Sethian circuit is sketched in figure 5.8.

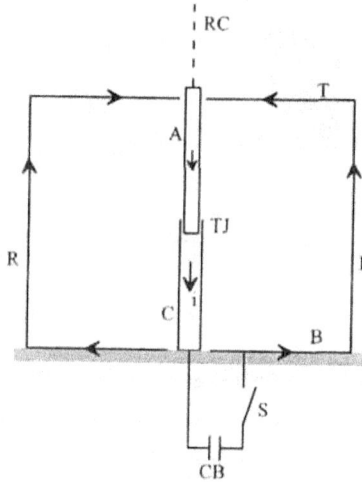

Figure 5.8 : Robson and Sethian coaxial armature experiment

The experiment shown in figure 5.8 concerned the axial motion of an armature tube, A, arranged coaxially with twelve equally spaced peripheral rods, R, on a circle. The rods were held by the top plate, T, and the bottom plate, B. The armature was hanging on a rubber cord, RC, and by means of the telescopic joint, TJ, it could slide into the central conductor tube, C. A large hole was cut in the center of the bottom plate through which C was connected to one of the poles of a capacitor bank, CB. The other pole of the capacitors was connected to the inner edge of the bottom plate.

When the switch, S, was closed, the capacitor would discharge and current would flow radially outward in the bottom plate, B, then up the peripheral conductor rods, R, and radially inward in the top plate, T, to the armature, A, and down to the central conductor, C. The capacitor voltage was high enough to ensure the formation of air arcs in the telescopic joint, TJ, and in the gap between the armature and the top plate. In figure 5.8, i denotes the total current flowing in the circuit.

Robson and Sethian then proceeded to calculate the Ampère forces which the various parts of the circuit should exert on the armature A. Let us denote the respective force components as follows:

$F_{T,A}$ - due to the top plate T;

$F_{C,A}$ - due to the central conductor C;

$F_{B,A}$ - due to the bottom plate B;

$F_{R,A}$ - due to the peripheral rods R.

Because of the coaxial geometry, all the force components had a net effect in the axial direction, either up or down. The algebraic sum of the components was found to be

$$F_{T,A} + F_{C,A} + F_{B,A} + F_{R,A} = 0 \qquad (5.7)$$

The authors checked the calculations and obtained the same result. In fact, using any geometry Ampère's force law always predicts a zero net longitudinal force on the armature. However as discussed in Chapter 2, calculation of the net force on an object as a result of a Newtonian force law is not the correct way of predicting internal stresses or relative accelerations. In the Robson and Sethian experiment, however, even correct application of the Newtonian Ampère force law predicts zero acceleration of the armature. This was found to be in complete agreement with experiment. The armature revealed no significant up or down displacement with currents as high as i=120 kA. Their paper did not explain why they ever expected to see a net longitudinal force, and why they thought this experiment was a test of the railgun recoil action.

The experiment of figure 5.8, however once more, affords an opportunity to explain where the real difference between the Ampère and Lorentz force laws lies. In the course of the Ampère force calculations it becomes clear that the downward $F_{T,A}$ force component on the armature has its seat quite close to the top of the armature. In contrast to this, the upward force component, $F_{C,A}$, resides near the bottom of the armature. These are the principal forces on the armature and they cause axial compression of the armature, without adding to the net force. No such compression is predicted by the Lorentz force law because Lorentz forces always act transverse to the current.

Robson and Sethian did not calculate the Ampère tension which arises in the armature due to the repulsion of current elements inside this conductor. This tension also does not add to the net force on the armature, and it would be zero in the case of the Lorentz force law. Electrodynamically generated compressive and tensile stresses in the metal lattice clearly distinguish the Newtonian electrodynamics from field theory.

In the Robson-Sethian experiment the maximum current was 120 kA. Using their force calculations, this comes to a compression force of 3100 N. It could certainly be detected with strain gauge measurements if the heat pulse would not make the measurement virtually impossible. The armature copper tube of 2.22 cm outside diameter, 0.7 mm wall thickness, and 12.7 cm length was far too strong to buckle in the brief period of the 100 μs current pulse. To show the Ampère compression, the authors found it necessary to employ conductors of much less buckling strength (flexural rigidity) as used in the experiment shown in figures 5.6 and 5.7.

In some of their tests Robson and Sethian [2.26] did, in fact, replace the armature of figure 5.8 by a 1 mm diameter aluminum wire. With this arrangement they found that the wire fragmented in Ampère tension, as discussed at length in Chapter 2, and the individual wire pieces were "bent into arcs of circles". In other words, they actually saw the effect of concurrent Ampère tension and wire buckling. Uncertain of their ground, Robson and Sethian briefly tried to explain away the unmistakable signs of Ampère stresses with unsubstantiated, and metallurgically wrong, thermal arguments. A similar suggestion was made by Ternan [2.11]. This has been fully refuted in Chapter 2.

Without the telescopic joint in figure 5.8, the Robson-Sethian experiment could be considered to be some form of railgun recoil test. Let us assume the telescopic joint was simply welded up. The upward force $F_{C.A} + F_{B.A} + F_{R.A}$ would then appear as tension in the weld. $F_{T.A}$ would compress the armature and the compression could be demonstrated by replacing the tubular armature with a thin strip, like the one used in the recoil experiment of figure 5.6. The tensile strength of the strip has to be such that it does not fragment due to Ampère tension. Then the conditions of the railgun recoil experiment of figure 5.6 would be approached. Robson and Sethian also ignored the relevance of the impulse pendulum experiment (described in Chapter 2) to the railgun recoil mechanism.

They overlooked another railgun recoil experiment published in 1982 [5.12]. This involved liquid mercury conductors. It therefore displayed, in addition to the Ampère recoil force, the electromagnetic jets observed in the straight-through mercury channel experiment of Chapter 2. The experimental setup is shown in figure 5.9. The tests were performed with direct current of a few hundred amperes. A car battery could have been employed as current source, making it easy for almost anyone to demonstrate the existence of longitudinal Ampère forces.

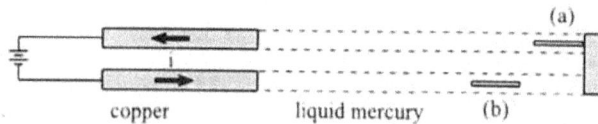

Figure 5.9 : Railgun recoil experiment with liquid mercury conductors

With respect to figure 5.9, rails of 45.7 cm length and ½" square cross-section were spaced ½" apart. The first 30.5 cm of each rail consisted of a solid copper bar and the remainder, up to the armature branch, was liquid mercury contained in rectangular grooves which had been milled into a thick plastic board. The fixed armature was a ½" square section copper bar set into the same plastic board. The armature fully bridged the mercury rails and the gap between them. At the breech end, the rails were connected to a power supply with which DC current up to 450 A could be passed through the circuit.

Right circular copper cylinders of 5 cm length and several different diameters, tinned all over for easy amalgamation with mercury, were placed on the mercury surfaces with one end of each cylinder touching the armature, as indicated by (a) in figure 5.9. While they were lying on the surface, the tinned copper rods were found to stick quite firmly to the mercury because of surface tension effects. But above a certain current value the rods would submerge and then much of the resistance to motion in the mercury apparently disappeared.

Submersion was the result of current sharing between the copper rod and the surrounding liquid mercury. This gave rise to lateral attraction between parallel metal filaments carrying current in the same direction. Due to the low resistivity of copper, the current in the copper rod was about half the total current in the trough section of figure 5.10, and it would therefore be urged toward the center of the mercury conductor. In the diagram the centering force has been denoted by F_c. In the steady state it will be in equilibrium with the buoyancy force, F_b, and the transverse electrodynamic force, F_t, due to repulsion by the current in the second rail.

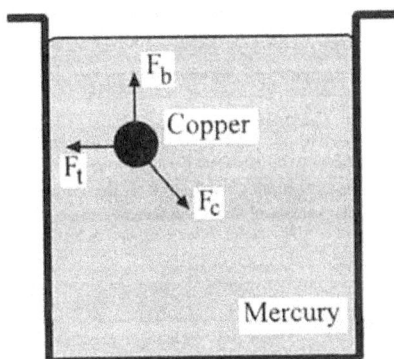

Figure 5.10 : Transverse forces acting on the copper cylinder while it is submerged in the mercury trough shown in figure 5.9

Experiments were carried out with three rod diameters of 0.300, 0.196 and 0.127 cm. The corresponding submersion currents were 280, 240 and 140 A. On switching the current off one second or less after submersion, the copper rods would surface some distance away from the armature branch, as shown by (b) in figure 5.9. The distance through which the rods travelled along the mercury rails was a function of rail current, cylinder diameter and time of current flow. With 0.196 cm diameter rods, longitudinal displacements of the order of two to three centimeters were observed, after having forced a current of 450 A to flow for less than one second. The longitudinal displacement is believed to have been caused by the railgun recoil force. Because of the large difference in the electrical conductivity between copper and liquid mercury, there can be little doubt that almost all the current in the copper cylinders was parallel to the rail axes. Hence the displacement of the cylinders could not have been caused by Lorentz forces on the cylinders.

For additional proof of the longitudinal recoil force, a further experiment was performed using the apparatus shown in figure 5.9. One of the copper rods was hand-held in the mercury surface with a pair of rubber handled tweezers and pressed with one end against the fixed armature, as indicated in figure 5.11. A pronounced jet of mercury could be seen to stream away from the other end of the rod, showing the strong longitudinal repulsion of mercury atoms from the copper rod. But this repulsion could not prevent the rod from pushing away from the armature as soon as the tweezer grip was released. From this it may be concluded that, unlike in Robson and Sethian's armature, the copper rods were able to accelerate longitudinally because the liquid mercury atoms could yield to repulsion, which was impossible with the solid center conductor C of figure 5.8.

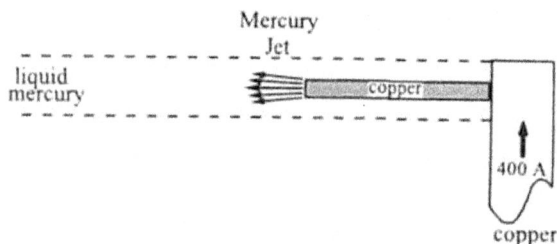

Figure 5.11 Liquid mercury jet at the end of the copper rod, held fixed in the surface of the liquid mercury trough

The Motionally Induced E.M.F.

In the direct conversion of electrical energy to mechanical work, or kinetic energy, the electrodynamic forces must be associated with electromotive forces. The only way in which the source of electrical power can expend energy, besides Joule heating, is by driving the current against an opposing back-electromotive-force, commonly referred to as a back-e.m.f.

The back-e.m.f., e_b, in the railgun circuit of figure 5.1 may be calculated with the aid of the selfinductance, L, of the rectangular circuit ABCD. Neumann provided Eq.1.44 for the induced e.m.f., e_n, in a circuit, n, due to the current, i_m, in another circuit, m, by making use of his concept of mutual inductance, $M_{m,n}$. This formula can be adapted to a single circuit of selfinductance, L, carrying current, i. The result is

$$e_b = - \frac{d}{dt} (L \, i) \qquad (5.8)$$

Both L and i may be variable, which leads to the partial differential equation

$$e_b = - L \frac{\partial i}{\partial t} - i \frac{\partial L}{\partial t} \qquad (5.9)$$

Furthermore, if the armature travels in the positive x-direction, Eq.5.9 may be written

$$e_b = - L \frac{\partial i}{\partial t} - i \frac{\partial L}{\partial x} \frac{\partial x}{\partial t} \qquad (5.10)$$

To show up the motionally induced effect, we consider the case of constant current and recognize that the armature velocity is $v = \partial x / \partial t$, so that

$$e_b = - i v \frac{\partial L}{\partial x} \tag{5.11}$$

It is with Eqs.5.10 and 5.11 of Neumann's theory of induction that the back-e.m.f. in railguns is normally estimated. Results obtained in this way are in good agreement with the measured back-e.m.f.

The selfinductance, L, may be calculated using the methods outlined in Chapter 3, starting, for example, with Eq.3.46. Rails, armature and breech then have to be resolved into g parallel filament circuits. Maxwell's geometric-mean-distance (GMD) may also be employed for the rectangular circuit.

To avoid the mathematical complexities of inductance calculations, the static selfinductance is frequently measured as a function of x, and the gradient $\partial L / \partial x$ is derived from a plot of L versus x. Any selfinductance measurement, of course, involves the induced back-e.m.f., and thus it is not surprising, therefore, that this procedure ultimately gives the correct back-e.m.f. result.

Strangely, field theory does not provide a method for correctly estimating the motionally induced back-e.m.f. The following analysis will demonstrate this. In field theory, the self and mutual inductances of circuits are defined by the magnetic flux linkage ϕ. Faraday's law should then predict the induced e.m.f. in terms of the rate of change of flux linkage. For the railgun this becomes

$$e_b = - \frac{d\phi}{dt} \tag{5.12}$$

Let B stand for the flux density due to the current, i. In the plane of ABCD in figure 5.1, the flux density is always perpendicular to the plane. If S denotes the area of the rectangle and ℓ the length of its periphery, the rate of change of flux linkage may be written

$$\frac{d\phi}{dt} = \oint_{\ell} (\vec{v} \times \vec{B}) \cdot d\vec{\ell} - \iint_{S} \left(\frac{\partial \vec{B}}{\partial t} \right) \cdot d\vec{S} \tag{5.13}$$

In other words, the rate of change of flux linkage can be expressed by two terms. The first term takes the current to be constant and accounts for the flux linkage change due to the motion of the circuit boundary. The second term assumes $v = 0$ and then computes the change in flux linkage due to the variation of current.

For a railgun operating under DC or quasi-DC conditions, Eqs.5.12 and 5.13 lead to

$$e_b = - \oint_{\ell} (\vec{v} \times \vec{B}) \cdot d\vec{\ell} \tag{5.14}$$

Since only the armature travels with velocity v and the rest of the railgun circuit is stationary, Eq.5.14 may be written

$$e_{AB} = - \int_{A}^{B} (\vec{v} \times \vec{B}) \cdot d\vec{l} \qquad (5.15)$$

According to field theory this is the motionally induced back-e.m.f. of the entire railgun circuit. No part of the induced voltage is found in the rails or the breech. Since Eq.5.11 depends on the selfinductance of the whole railgun circuit, we should suspect that there is disagreement between Eqs.5.11 and 5.15. Measurements readily confirm that the induced back-e.m.f. is twice as large as predicted by the field theory formula, Eq.5.15.

The inadequacy of Faraday's law has been revealed by yet another railgun measurement. This has sometimes been called the 'muzzle voltage puzzle'. It becomes apparent when a voltmeter is connected across the gun muzzle between E and F of figure 5.1.

Unlike the rails, which can be assumed to have negligible resistance, the armature is known to have a high resistance, mostly due to the sliding contacts with the rails. Let R_{AB} be the resistance of the armature giving a volt-drop (i R_{AB}) across it. Applying Kirchhoff's voltage law (the algebraic sum of all voltages in a closed circuit is zero) to the circuit ABFE and calling the muzzle voltage V_m, we must have

$$V_m - iR_{AB} - e_{AB} = 0 \qquad (5.16)$$

The implication of Kirchhoff's law is that the muzzle voltage should increase as the armature accelerates along the rails because v increases and this is contained in e_{AB}. Measurements show no velocity dependent increase in the muzzle voltage. Typical oscilloscope traces of V_m which were obtained by Stainsby and Bedford [5.15], are reproduced in figure 5.12. In these measurements the armature was a travelling electric arc in air between the rails. This explains the large voltage drop of 200 V across the armature. The arc pushed against a plastic projectile and accelerated it for the first 0.9 ms of figure 5.12. Then a sudden large increase in muzzle voltage occurred due to the arc leaving the railgun and dissipating the remaining energy stored in the gun inductance by what is known as the muzzle flash.

The velocity dependent part, e_{AB}, of Eq.5.16, should have become prominent in the second half of the 0.9 ms record of figure 5.12, but there is no indication of it. Most investigators now appear to be resigned to the absence of e_{AB} in the muzzle voltage measurement. For example, in the description of one of the most powerful railguns built to date, Holland et al [5.16] simply state: "A voltage probe located at the railgun muzzle measures the arc-voltage drop as a function of time."

Equation 5.16 assumes that there are no e.m.f.s induced in the rail portions AE and BF and in the voltmeter branch EF. These conductors do not carry current. Consider an element in the rail portion AE. From Eq.5.15 it is obvious that field theory obtains the motionally induced e.m.f. with the help of a motionally induced electric field strength.

$(\vec{v} \times \vec{B})$. Special relativity requires that $\vec{v}$ is the relative velocity between the moving charge and "the observer". In the case considered here, the observer is the muzzle voltmeter which is stationary with respect to the rails. In AE the magnetic flux density is perpendicular to the railgun plane and the only charge velocity in the rail would be the drift velocity of the conduction electrons. However, when no current flows, field theory assumes that this drift velocity is zero, i.e. $\vec{v} = 0$ in Eq.5.15. Hence, according to relativistic field theory, no e.m.f.'s can be induced in the current-free branches AE, BF and EF.

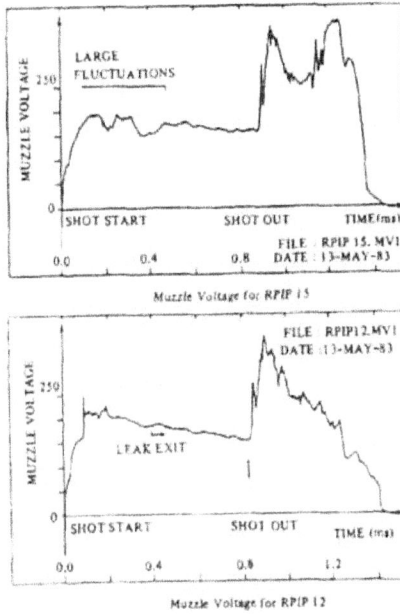

Figure 5.12 : Muzzle voltage measurements by Stainsby and Bedford [5.15]

In the active rail branches, AD and BC, there is current flowing and the conduction electrons move with a finite drift velocity $\vec{v}$ with respect to the stationary observer, in this case the breech voltmeter. But now it is found that, with a vertical magnetic flux density and an axial charge motion, the vector product of Eq.5.14 gives an induced electric field which is transverse to the conductor axis. This cannot be detected by the V_b voltmeter. Relativistic field theory, therefore, suggests that no e.m.f.s can be induced in the conductor branches AD, BC and DC.

The special theory of relativity only permits the induction of a back-e.m.f. in the armature. The relevant component of $\vec{v}$ is the charge velocity in the direction of the armature velocity with respect to the stationary observers V_b and V_m. This represents convection of the conduction electrons with the armature metal, and it is not the electron drift velocity due to current flow.

In summary, relativistic field theory underestimates the total induced back-e.m.f. in the railgun and it is unable to explain why the muzzle voltage is not a function of the armature velocity. Both of these problems have been resolved with Neumann's law of induction [5.17]. This problem deals exclusively with motionally induced e.m.f 's expressed by Eq.1.39. It is important to remember that Neumann's law does not stand for a reciprocal interaction between elements, but for a one-way action of cause and effect. The cause is a current element and the effect is an induced e.m.f. in a conductor element which may, or may not, carry current.

Let us apply Eq.1.39 to the three elements dm, dn and $d\ell$ of the railgun circuit of figure 5.13, of which only the first two elements carry the current i. Without current, element $d\ell$ cannot cause an e.m.f. in dm, yet $i_m$dm will induce $\Delta e_{\ell,m}$ in $d\ell$ .With this notation, the other non-zero e.m.f.s amongst the three elements are $\Delta e_{m,n}$ and $\Delta e_{n,m}$.

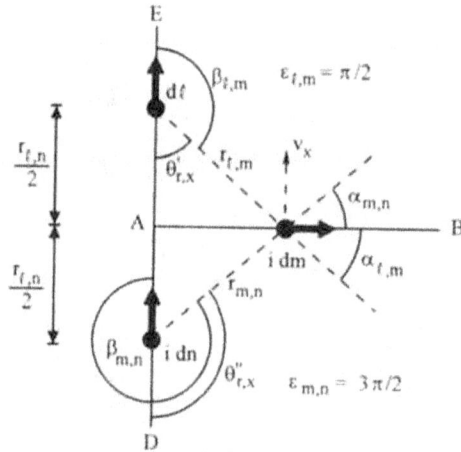

Figure 5.13 : Inductive interactions of three conductor elements of the railgun

With Ampère's force law Eq.1.24 substituted into Neumann's law of induction, Eq.1.39, we obtain, for the dm-dn pair

$$\Delta e_{m,n} = - i \frac{dm\ dn}{r_{m,n}^2} (2\cos \varepsilon_{m,n} - 3 \cos \alpha_{m,n} \cos \beta_{m,n}) v_x \cos \theta_{r,x}'' \qquad (5.17)$$

All the parameters of this equation are shown in figure 5.13. In the case of Eq.5.17 it is found that $\varepsilon_{m,n} = 3\pi/2$, $\cos\beta_{m,n} = \sin\alpha_{m,n} = - \cos\theta_{r,x}''$. Therefore

$$\Delta e_{m,n} = - 3 i \frac{dm\ dn}{r_{m,n}^2} v_x \cos \alpha_{m,n} \sin^2\alpha_{m,n} \qquad (5.18)$$

All the quantities in Neumann's law, Eq.1.39, are the same for $\Delta e_{m,n}$ and $\Delta e_{n,m}$. If we call Δe our unit of e.m.f., then

$$\Delta e_{m,n} = \Delta e_{n,m} = -\Delta e \qquad (5.19)$$

Both e.m.f.s are negative, that is opposed to the current, as is appropriate for back-e.m.f.s.
The third of the motionally induced e.m.f.s is

$$\Delta e_{\ell,m} = -i \frac{dm \; d\ell}{r_{\ell,m}^2} (2 \cos \varepsilon_{\ell,m} - 3 \cos \alpha_{\ell,m} \cos \beta_{\ell,m}) v_x \cos \theta'_{r,x} \qquad (5.20)$$

For the $d\ell$-dm combination $\varepsilon_{\ell,m} = \pi/2$, $\cos\beta_{\ell,m} = -\sin\alpha_{\ell,m} = -\cos\theta_{r,x}$, $r_{\ell,m} = r_{m,n}$, $\alpha_{\ell,m} = \alpha_{m,n}$ and $d\ell = dn$, so that Eq.5.20 may be written

$$\Delta e_{\ell,m} = -3i \frac{dm \; dn}{r_{m,n}^2} v_x \cos \alpha_{m,n} \sin^2 \alpha_{m,n} \qquad (5.21)$$

Hence

$$\Delta e_{m,n} = \Delta e_{n,m} = \Delta e_{\ell,m} = -\Delta e \qquad (5.22)$$

In contradiction to relativistic field theory, Neumann's law places motionally induced e.m.f.s in the rails, not only in the current carrying portions, but also in the sections ahead of the armature. From the fact that $\Delta e_{\ell,m} = \Delta e_{m,n}$ oppose each other in the loop of the muzzle voltmeter, we can foresee the possibility of this voltmeter not registering the motionally induced e.m.f. in the armature.

A finite element analysis was performed on the whole of the railgun circuit of figure 5.1. After summing the elemental induced e.m.f.s over the various circuit branches, the results depicted in figure 5.14 were obtained. The rail e.m.f.s, like the Ampère forces, were found to be sharply concentrated near the armature junctions. This means that the voltmeter at the muzzle sees no change until the armature comes very close to the end of the rails.

It can now be understood why the net induced e.m.f. in the muzzle voltmeter loop is (2e-2e=0), in conformity with experimental findings. Furthermore, the total motionally induced e.m.f. in ABCD is (2e+2e= 4e), or twice as large as predicted by field theory. This also agrees with experiment.

In an exchange of published letters, Allen [5.18, 5.19, 5.20] has tried to defend field theory. One of his claims is that in Eq.5.13, $\partial B/\partial t$ is not equal to zero, even when the current is constant because, according to him, "the magnetic field at AB increases from zero as the projectile moves past."

The partial differential in Eq.5.13, however, refers to the change in flux linkage with the metallic circuit ABCD while no displacement of any part of the circuit takes place. The other variable, x, is frozen for the purpose of the partial differentiation. It is plainly wrong to speak of the projectile moving past AB while considering the partial differential $\partial B/\partial t$.

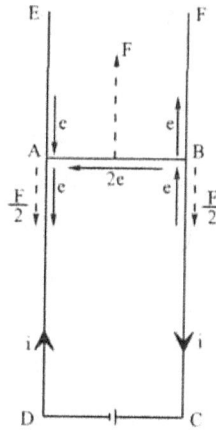

Figure 5.14   Computed motionally induced e.m.f.s in the railgun

We conclude this section on the motionally induced e.m.f. with an argument put forward by an irate anonymous reviewer. He claimed that $\vec{v}$ in Eq.5.15 was the velocity of the armature relative to the local magnetic flux density, $\vec{B}$, and that this flux density moves with the armature. Hence $\vec{v}$ should be zero and with it would disappear the lone induced e.m.f. component given by field theory. This, however, violates the observer rule of special relativity. Having done this, the reviewer fell back on Eq.5.11 for the motionally induced railgun e.m.f. Little did he know that this was precisely what the authors had in mind - induction must be action at a distance!

The motionally induced railgun e.m.f. seems to be the first example that has come to light in which Faraday's law and Maxwell's equations have failed. Up to this point only the Lorentz force law was in trouble, and Maxwell held that the choice of force law had nothing to do with his equations. All these shortcomings of relativistic electromagnetism make the return of the Newtonian electrodynamics in the twenty-first century virtually inevitable.

To close this chapter we mention a strange feature of railgun behavior which, at the time of writing, has yet to find an explanation in relativistic or Newtonian electromagnetism. It concerns an accidental discovery made in December 1991 in the Oxford laboratory of Neal Graneau.

While working on the experiment with the screened enclosure, depicted by figure 5.2, the authors accidentally placed a ferromagnetic carbon steel rod armature on the rails and observed that it rolled in the wrong direction toward the battery. This surprising behavior was perfectly repeatable. Because of the pressure of other research it was not published. However in May of 1992 Peter Graneau was asked to review a paper by the Hungarian authors

Bardocz-Todor and Meszaros [5.21] who had discovered the same effect. PG recommended the paper for publication, but requested linguistic improvements. As far as we know, the paper by the Hungarian authors has not been published. If this is correct, then our report here is the first public mention of a railgun armature being driven toward the current source. While not yet understood, the importance of this discovery cannot be over-stressed.

Chapter 5 References

5.1   A. Egeland,"Birkeland's electromagnetic gun: A historical review", IEEE Transactions on Plasma Science, Vol.17, p.73, 1989.

5.2   S.C. Rashleigh, R.A. Marshall, "Electromagnetic acceleration of macroparticles to high velocities", Journal of Applied Physics, Vol.49, p.2540, 1978.

5.3   R.P. Feynman, R.B. Leighton, M. Sands, *The Feynman lectures on physics*, Vol.2, p.27.8-27.9, Addison-Wesley, Reading MA, 1964.

5.4   D.W. Deis, D.W. Scherbarth, G.L. Ferrentino, "EMACK electromagnetic launcher commissioning", IEEE Transactions on Magnetics, Vol. MAG-20, p.245, 1984.

5.5   P. Graneau, "Amperian recoil and the efficiency of railguns", Journal of Applied Physics, Vol.62, p.3006, 1987.

5.6   J.E. Allen, T.V. Jones, "Relativistic recoil and the railgun", Journal of Applied Physics, Vol.67, p.18, 1990.

5.7   J.H. Poynting, *Collected scientific papers*, Cambridge University Press, Cambridge, 1920, p.182.

5.8   P. Graneau, "Nonlocal action in the induction motor", Foundations of Physics Letters, Vol.4, p.499, 1991.

5.9   J.S. Bell, *Speakable and unspeakable in quantum mechanics*, Cambridge University Press, New York, 1987.

5.10   A. Aspect, J. Dalibard, G. Roger, "Experimental test of Bell's inequalities using time-varying analyzers", Physical Review Letters, Vol.49, p.1804, 1982.

5.11   J.P. Vigier, University Pierre et Marie Curie, Paris, private communication.

5.12   P. Graneau, "Application of Ampère's force law to railgun accelerators", Journal of Applied Physics, Vol.53, p.6648, 1982.

5.13   P. Graneau, "Railgun recoil and relativity", Journal of Physics D: Applied Physics, Vol.20, p.391, 1987.

5.14   W.F. Weldon, M.D. Driga, H.H. Woodson, "Recoil in electromagnetic railguns", IEEE Transactions on Magnetics, Vol. MAG-22, p.1808, 1986.

5.15   D.F. Stainsby, A.J. Bedford, "Some diagnostic interpretations from railgun plasma profile experiments", IEEE Transactions on Magnetics, Vol. MAG-20, p.323, 1984.

5.16   M.M. Holland, G.M. Wilkinson, A.P. Krickhuhn, R. Dethlefsen, "Six megajoule railgun facility", IEEE Transactions on Magnetics, Vol. MAG-22, p.1521, 1986.

5.17   P. Graneau, D.S. Thompson, S.L. Morrill, "The motionally induced back-e.m.f. in railguns", Physics Letters A, Vol.145, p.396, 1990.

5.18   J.E. Allen, "The motionally induced back-e.m.f. in railguns", Physics Letters A, Vol.150, p.151, 1990.

5.19   P. Graneau, "Comment on 'The motionally induced back-e.m.f. in railguns'", Physics Letters A, Vol.160, p.490, 1991.

5.20   J.E. Allen, "Reply to Comment on 'The motionally induced back-e.m.f. in railguns'", Physics Letters A, Vol.160, p.492, 1991.

5.21   A. Bardocz, M. Meszaros, "A new effect of electrodynamics?", Department of Physics, Berzsenyi Daniel College and Institute of Physics of Budapest University of Technology, manuscript dated March, 1992.

Electrodynamics of Arc Explosions

Conventional Arc Physics

In this book the discussion of arcs is limited to high-density arcs, ranging from the density of atmospheric air to that of liquids and particularly water. The experimental evidence presented in this chapter, and available in 1995, leaves little doubt that high-density arcs are subject to the Newtonian electrodynamics. On the other hand, we know that isolated charges drifting in vacuum are governed by the very different electrodynamic laws of relativistic electromagnetism. Somewhere between the high density plasma state and isolated charges travelling in evacuated particle accelerators, a transition must occur from the electrodynamics of dense matter to the Lorentz force regime of independent charged particles. The pressure range in which this transition takes place has not yet been determined.

In a dense gaseous medium one observes sparks and arcs. Here we define sparks as low current discharges in fluid media. The spark indicates the breakdown of the dielectric fluid by the formation of electron avalanches, streamers, and thus forming a thin plasma filament. The rapid loss of ions from the thin filament makes the ion replacement by electron avalanches a continuous necessity. This process requires a large driving voltage. Depending on filament lengths, spark volt-drops are measured in kilo-volts. Only voltage sources with internal or external impedance will permit the formation of sparks, or the violently fluctuating sparking phenomenon. To prevent sparks from becoming arcs, the current has to be limited to a fraction of one ampere.

If the current is allowed to reach several amperes, the spark develops into an arc of lower volt-drop, frequently less than 100 V. This indicates the existence of an established plasma channel which acts much like an ordinary metallic conductor. The arc current is almost entirely due to the flow of electrons. Positive ions are continuously lost to recombination at the electrodes and the fluid environment. These ions have to be replaced by ionizing collisions of electrons with neutral atoms or molecules. Ionization accounts for part of the arc voltage drop. The energy absorbed by ionizing collisions is stored by the ions and later recovered as heat when the ions recombine with electrons. In many instances a major part of the arc voltage generates ohmic heat by electron-ion collisions. This gives rise to the arc resistance.

If the electrodynamic forces developed in the arc do mechanical work on the plasma or its environment, then a back-electromotive-force (e.m.f.) must be induced in the arc column. The energy expended by the power source in driving the current against the back-e.m.f. does not produce heat, but is directly converted to mechanical work. The same energy conversion mechanism drives electric motors.

The rate of energy expenditure (power) of the current, i, in overcoming the back-e.m.f., e_b, is e_b i (watts). For example, if an electrodynamic force, F, pushes the surrounding medium outward with the velocity v, energy conservation requires that

$$F v = e_b i \qquad (6.1)$$

If the arc column has a resistance R, then the resistive arc voltage drop iR, and the induced back-e.m.f. e_b, act in the same direction, both opposing the flow of i.

A practical consequence of this is that ordinary arc voltage measurements cannot distinguish between the energy expenditure on ionization, electron-ion collisions and mechanical work done by the arc. This has confused the search for an explanation of the considerable forces which drive pulsed arc explosions. Ohmic heating increases the plasma temperature and pressure. However most investigators, unaware of the Ampère forces, have considered only the thermal mechanism of causing arc explosions. Back-e.m.f.s, if they were present, would have been treated as resistive volt drops which were supposed to increase the plasma temperature and pressure. There existed no incentive to look for the conversion of magnetic energy to mechanical work, because the Lorentz force acts as inward pinch forces, which would try to decrease the arc diameter, and this is not observed.

It might be thought that a way of separating the back-e.m.f. from the resistive volt drop would be to prevent the electrodynamic forces in the plasma from doing mechanical work, as would be the case in a very rigid arc cavity. This would minimize the storage of strain and kinetic energy in the arc environment and should result in a decrease or elimination of the back-e.m.f. It seems possible, however, that the arc forces would then cause chaotic motion of the plasma ions which would increase the plasma temperature and manifest itself as an increase in the arc resistance.

The arc plasma is unlikely to do work on itself by internal compression under the action of pinch forces. The very existence of the outward explosions proves that the pinch action is overcome by expansive forces.

If we ignore the attention lightning discharges have attracted for the past few centuries, arc research is about one hundred years old. The extensive literature on the subject concerns itself with such topics as: ignition, arc maintenance, extinction, temperature, voltage distribution, ion density and electrode phenomena. The books by Cobine [6.1] and Hoyaux [6.2] present typical conventional arc physics. From the point of view of arc explosions, the most pertinent conventional topic is arc temperature. The very highest temperatures are measured by spectroscopic means, and are of the order of 40,000°K. They apply to atmospheric arcs of lightning strength.

The issue of electrodynamic forces in arcs is hardly ever mentioned. The reason must be that conventional electromagnetism ascribes electrodynamic forces to the Lorentz mechanism. With the common assumption of overall electrical neutrality of the arc plasma, only the magnetic component of the Lorentz force can affect arc behavior. This component is

$$\vec{F} = q\,\vec{v}\times\vec{B} \qquad\qquad (6.2)$$

where q is the charge of the test particle, $\vec{v}$ the velocity of the particle relative to the observer, and $\vec{B}$ is the local magnetic flux density at the charge location. When the observer measuring the explosion forces is stationary in the laboratory, the relative velocity is in the direction of current flow. The flux density is azimuthal, and therefore, the force F acts radially inward. Hence the most important consequence of Eq.6.2 is the magnetic pinch of the arc column. On account of the inverse square of the distance relationship, which is concealed by Eq.6.2, the transverse force on the arc column, exerted by remote parts of the circuit, is far smaller than the pinch force. Equation 6.2 gives no longitudinal forces in the arc column. Nonetheless, since the pinch force is exerted on a fluid medium, it will be converted hydrostatically to an axial thrust on the electrodes. The axial thrust, however large, cannot lead to a radial outward explosion, because when the outward forces become equal to the pinch forces, the axial thrust ceases to exist. An analogy is the compression of gas in a cylinder by a piston. The reaction thrust on the piston can never exceed the force applied to the piston. Anyway, Aspden [6.3] has pointed out that the pinch force is about four orders of magnitude smaller than the forces measured in water arc explosions.

Finally, it should be remembered that the Lorentz current element is the moving charge multiplied by its velocity and not the relatively slow ion of the Ampère electrodynamics. The radial Lorentz pinch pressure must then arise from Coulomb forces between the electron concentration along the core of the arc column and the excess positive ions in the outer region of the arc. In contrast, the Ampère pinch forces are applied to the positive ions without assistance from Coulomb forces.

Transient High Current Arcs

When a welding arc is first struck it produces the sound of a gun shot. Later, when it is burning steadily, the sound can best be described as hissing. The acoustic behavior of the arc indicates that its ignition is responsible for an explosive shock wave in the atmosphere which then settles down to fluctuating air pressure. Judging by the shock intensity, there is no doubt that the strength of the explosion is related to the peak value of the current pulse.

In the first half of the present century the flow of 1000 A in a welding arc was considered to be a large current. Recent decades have seen the growth of pulse current amplitudes to one mega-ampere and more. Very large current pulses are required for driving railguns, thermonuclear fusion reactors and the simulation of electromagnetic pulses (EMP) of atomic explosions. Extrapolating the Ampère tension graph of figure 2.1 to one mega-ampere suggests a force of 450 kN. This would be sufficient to pull apart an aluminum conductor seven centimeters in diameter!

Both the Joule heating and the electrodynamic forces in the arc are proportional to the square of the current. It is not surprising, therefore, that mega-ampere pulses result in powerful arc explosions. It is the cost of energy storage facilities which mitigates against the maintenance of high current, but for a very brief period of time. This is the reason why high current arcs have all been transient events.

The best known example of a transient high current arc is lightning. Thunder accompanying lightning strokes provides ample proof of the explosive nature of these naturally occurring arcs. It is very difficult to measure the pressure pulse and other parameters of the lightning discharge because of its unpredictable location and timing. But short arcs, carrying the same current as lightning, can be struck at will in the laboratory. Capacitor banks usually furnish the energy for pulsed arc experiments.

A laboratory technique has been developed for measuring forces and pressure in transient high current arcs. The dynamic force measurement turns out to be easier than force determinations on continuously burning arcs. The reason is that the pulsed arc generates a limited amount of heat. This allows the arc to be totally enclosed by strong walls. Steadily burning arcs, by contrast, melt the electrodes and make great demands on the heat resistance of the dielectric enclosure.

Many of the arc experiments, described in this chapter, were performed in an MIT electrodynamics laboratory. Normally, the arc cavity was formed by setting strong metal electrodes into a substantial block of glass-fibre reinforced epoxy. The upper wall of the arc cavity was a thin dielectric plate laid on the epoxy block. A metal weight was then placed on the plate. The explosion would throw this weight vertically up into the air. It was easy to measure the throw-height of the weight. From this measurement the time-average force could be calculated in the following way.

Let the weight W have a mass m. If the throw-height is h, and the acceleration due to gravity g, then the potential energy E_p acquired by the weight at the apogee of its trajectory is

$$E_p = mgh \qquad (6.3)$$

With a sufficiently massive weight, the acceleration is slow enough that the force can be considered to be impulsive, which assumes that the force may have ceased before the mass acquires any significant displacement. This assumption allows us to equate the potential energy, E_p, to the initial kinetic energy E_k of the weight, given by

$$E_k = \frac{1}{2} m v_0^2 \qquad (6.4)$$

In Eq 6.4 v_0 is the initial velocity of W as it leaves the arc enclosure. Equating the potential and kinetic energies, the initial velocity is found to be

$$v_0 = \sqrt{2gh} \qquad (6.5)$$

Hence the measurement of h leads directly to v_0.

The laws of mechanics require that the initial momentum (mv_0), which was imparted to the weight W by the transient arc force F(t), must be equal to the impulse of the transient force. F generally varies with t, and therefore the impulse has to be written

$$\int_0^{\Delta t} F(t)\, dt \;=\; m\, v_0 \qquad\qquad (6.6)$$

In the last equation Δt is the current pulse duration. If <F> is the time-average of the arc force, then Eq.6.6 may be rewritten

$$<F> \;=\; \frac{m\, v_0}{\Delta t} \;=\; \frac{m\, \sqrt{2\,g\,h}}{\Delta t} \qquad\qquad (6.7)$$

Hence the measurement of m, h, and t allows the determination of the average arc force.

How do we know that the arc pressure pulse is no longer than the current pulse? For example, when dealing with a water arc, the current might generate high pressure steam, and this pressure could accelerate the weight for a longer time than the duration of the current pulse.

Several independent experimental proofs will be outlined in the remainder of this chapter which show that the arc explosions are not driven by steam or the expansion of hot gasses. At the same time experimental facts will be revealed which directly support the electrodynamic force mechanism. It is on the basis of this experimental evidence, and also with thermal analysis, that the use of the current pulse duration Δt in Eq.6.7 is justified.

The MIT experiments were all performed with the 8 μF, 100 kV, 40 kJ capacitor bank shown in figure 2.14 and described in conjunction with the impulse pendulum experiment of Chapter 2. The Rogowski coil surrounding the conductor of the discharge circuit, furnished a di/dt-signal which was integrated by a R-C network to convert it to a current (i) signal. This signal was displayed on an oscilloscope and photographed.

Unless unusual events occurred, the oscillograms were underdamped, exponentially decaying sinusoidal waves. The damping of the current oscillation is due primarily to the conductor resistance. Only if the resistance remains constant throughout the duration of the oscillation will the damping envelope be a truly exponential curve.

The ideal resistively damped current in the capacitor circuit is expressed by

$$i \;=\; I_0\, e^{t\,\tau}\, \sin(\omega\, t) \qquad\qquad (6.8)$$

where $\omega = 2\pi f$ and f is the ringing frequency. If Z is the circuit impedance and V_0 the voltage to which the capacitors were originally charged, the initial current amplitude becomes

$$I_0 \;=\; \frac{V_0}{Z} \qquad\qquad (6.9)$$

In many practical cases the resistance of the circuit is negligible compared to the reactance. The total impedance of the circuit is then approximately equal to the surge impedance

$$Z = \sqrt{\frac{L}{C}} \qquad (6.10)$$

where L is the circuit selfinductance and C the capacitance of the bank of capacitors.
T in Eq.6.8 is the damping time constant which is related to the circuit resistance by

$$T = \frac{2L}{R} \qquad (6.11)$$

Both the ohmic heat developed in the circuit and the electrodynamic forces generated by the current involve a quantity known as the action integral of the current pulse. This will be denoted by A. Its definition is

$$A = \int_0^\infty i^2 \, dt \qquad (6.12)$$

When the current is exponentially damped, as in Eq.6.8, we find the action integral becomes

$$A = I_0^2 \left[\left(\frac{T}{4} \right) - \left(\frac{(1/T)}{(2/T)^2 + (2\omega)^2} \right) \right] \qquad (6.13)$$

It was found that in all the reported arc explosions the second term of this equation was negligible relative to the first term. This led to the useful approximation

$$A \approx \frac{I_0^2 \, T}{4} \qquad (6.14)$$

A further simplification of the mathematical analysis of electrodynamic situations arises from the fact that the electrodynamic force on a circuit, or part of a circuit, can always be expressed by

$$F = \left(\frac{\mu_0}{4\pi} \right) k i^2 \qquad (6.15)$$

In this equation k is a dimensionless constant depending entirely on the geometry of the current path and the force law (Ampère's or Grassmann's). Equation (6.15) holds for relativistic electromagnetism as well as the Newtonian electrodynamics, but k is a different number in the two theories.

Every current pulse has a root-mean-square value which will be denoted by i_{rms}. If this value is substituted in Eq.6.15, the time average force <F> is obtained. For the damped current pulse of Eq.6.8, the root mean-square value would be zero because, theoretically, the pulse lasts forever. In practice the time constant can be used to describe the length of the pulse, for at the end of T the i²-amplitude is down to 13 percent of its original value. The approximate rms-current would then be

$$i_{rms} = \sqrt{\frac{\int_0^\infty i^2 dt}{T}} \cdot \frac{I_0}{2} \tag{6.16}$$

The electrodynamic force is a maximum when the current in Eq.6.15 is a maximum. This will occur at the first current peak of Eq.6.8. Hence the average and maximum forces become

$$<F> = \left(\frac{\mu_0}{4\pi}\right) k \left(\frac{I_0^2}{4}\right) \tag{6.17}$$

$$F_{max} = \left(\frac{\mu_0}{4\pi}\right) k\, i_{max}^2 \tag{6.18}$$

With this preamble we are ready to study a variety of arc explosions and also the forces developed by a few steadily burning arcs. The most interesting explosions so far investigated are those due to underwater arcs. They will be considered first.

Saltwater Cup Experiments

A major difference between electronic conduction through liquid metals and through plasma arcs is that the metal always presents an ionized state, whereas in the plasma arc some energy has to be expended to create the ions. Conduction electrons of the metal are so loosely bound to their atoms that thermal agitation is sufficient to set them free and leave behind a lattice of positive ions. The free electron gas, as it is sometimes called, makes for the high conductivity of metals. In the plasma, the ionization energy is stored (latent heat of arc formation) and later recovered as heat when the ions recombine with electrons. Most of the recombination takes place on the electrode faces.

Of course liquid metals are far denser than the arc plasmas of dielectric fluids. Electrodynamic forces in liquid metals meet more inertial resistance, and the explosions in them are more gentle than those of water and air arcs.

Compare 1 cm³ of water with 1 cm³ of liquid mercury. The density of liquid mercury is 13.6 times that of water. Let us assume that each of these two volumes receives an electrodynamic impulse of 5 Ns. Using Eq.6.6 and knowing that 1 cm³ of water weighs one gram, we find the water will acquire a velocity of 5000 m/s. The same calculation yields a mercury velocity of 368 m/s. The velocity is one way of describing the violence of the explosion. Another can be gleaned from the kinetic energies ($\frac{1}{2}mv^2$). For the mercury this comes to 919 J, while the water gains 12.5 kJ.

The saltwater cup was originally designed to demonstrate the liquid mercury fountain of figure 2.22. In figure 6.1 the same cup has been adapted for saltwater experiments [6.4]. It was filled with a saturated solution of common salt (NaCl) in tap water at room temperature. The power supply for these experiments was an energy storage capacitor C in series with a selfinductance L and a switch connected to the rod-and-ring electrodes in the cup. The capacitor was charged to the voltage V_0. When the switch was closed, and depending on the values of C and V_0, the discharge through the water was either silent, and left the water undisturbed, or it resulted in a luminous underwater arc in the vicinity of, or at the endface of the copper rod. Arcing was not a random event, but depended systematically on the capacitance and charging voltage. Visible arcs were always accompanied by a snapping sound and shock disturbance of the saltwater.

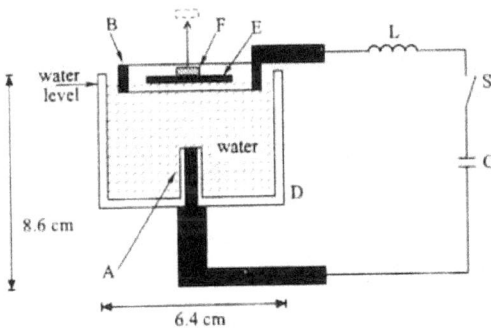

Figure 6.1 . Cross-section of saltwater cup, drawn to scale
A: insulated copper rod with bare endface. B: bare copper ring electrode,
D: dielectric cup containing saltwater, E: wooden disc float, F: 2.8 gm metal weight

With larger arc current pulses, a column of water was thrown up in the air directly above the copper rod. The explosive nature of the arc could best be demonstrated by placing a wooden float E on the water surface, inside the ring electrode, and setting a 2.8 gm metal weight F on the float. Then with a silent discharge the float and weight would remain still. But the formation of an arc would fling the weight upward by as much as 20 cm, without spilling any water. The float was behaving as if the mechanical impulse ceased as soon as the arc was extinguished. No follow-through push from expanding steam, nor any vapor escape from the cup, could be discerned.

The graph of figure 6.2 shows the combination of C and Vo values chosen for 17 experimental shots. It also indicates which shots resulted in arcs and which did not. An approximate boundary has been drawn between the arcing and silent discharge regions. Seventeen experimental points are insufficient to prove that this boundary is a straight line.

With the larger capacitances, lower voltages were able to initiate arcs. This suggests a certain number of ions, say Q, have to cross the gap before an arc can be formed. If this is true we should have

$$Q = C\,V_0 \sim \text{constant} \tag{6.19}$$

According to this argument the arcing boundary of figure 6.2 should be a hyperbola. This possibility is not excluded by the relatively few experimental points.

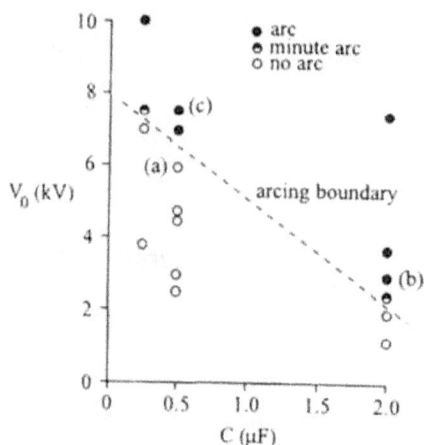

Figure 6.2 : Charging voltages and capacitances of saltwater cup experiments

Saltwater was chosen for this investigation in order to initiate the discharge the moment the switch in the capacitor circuit was closed. Pure water is a good insulator, which can be broken down with sufficiently high voltages, but usually gives an unpredictable statistical time lag between switch closing and arcing. The problem could be overcome with a thin wire bridging the gap between the electrodes. Metal atoms from the exploded wire would then mix with the water and may inhibit the strength of the explosion. In tap water, as opposed to distilled water, the discharge would still be delayed by random intervals. This behavior was found to be accentuated in distilled water when it would sometimes take a minute or more for the arc to strike. Higher voltages would shorten the delay time. It was observed that the strength of the water plasma explosions increased markedly when the saltwater was replaced by tap water or distilled water.

The loss of capacitor charge and energy during the delay period, before the arc was struck, was investigated in connection with electro-hydraulic metal forming. This is a technology which had its beginnings in the 1950s. The process relies on an underwater arc explosion pressing a metal plate into a mould. Gilchrist and Crossland [6.5] reported a series of measurements, and commented on the silent discharge delay as follows:

> "The loss of energy experienced in the delay period would suggest that the conductivity (ion concentration) of the discharge medium may be important and tests were carried out with tap water with varying amounts of sodium sulphate added to increase the conductivity from 73 μ mho/cm up to 850 μ mho/cm. This had no detectable effect on the length of the delay time, but the energy dissipated during the delay period was found to increase quite rapidly as the conductivity was raised. When distilled water was used as a discharge medium, breakdown frequently did not occur at a gap width of 5/8" or greater, and even when it did occur it was only after a long and variable delay time."

In the Gilchrist and Crossland measurements as much as thirty percent of the energy stored in the capacitors was dissipated in silent electrolytic pre-discharges in tap water. As the electrodynamic forces are proportional to the square of the current, this energy loss should have halved the explosion force. Unfortunately, these investigators did not determine the strength of the explosions.

From the saltwater cup experiments and the electro-hydraulic forming investigation it can be safely concluded that electrolytic currents are incapable of producing ponderomotive electrodynamic forces or in any way contribute to the arc plasma explosion. This is a surprising and scientifically important result because it is normally assumed that an electrolytic ion current is associated with the same magnetic effects as an electron current.

Another remarkable result of the saltwater cup experiments was that combinations of C and V_0 could be chosen so that, for the same amount of energy stored in the capacitor bank, the discharge was in one case silent and in the other case explosive. Presumably the energy was dissipated along the same current path and in approximately the same time. This alone almost rules out that the arc explosions were driven by thermodynamic forces. It was this finding which initiated the wider investigation of arc plasma forces.

Points (a) and (b) of figure 6.2 were chosen for further analysis in table 6.1. Both discharges were associated with the same stored energy of 9 J, but one resulted in an arc explosion and the other did not.

The purpose of the large inductance L= 876 μH in the discharge circuit (see figure 6.1) was to prolong the discharge for a more reliable observation of the current oscillations. The saltwater resistance between the electrodes was small enough to cause an underdamped oscillation in both the electrolytic and plasma conduction modes, as shown in figure 6.3.

Electrodynamics of Arc Explosions

Assuming the general form of the discharge current to be given by Eq.6.8, and the current amplitude I_0 and the damping time constant T by Eqs.6.9 and 6.11 respectively, it is possible to split the total circuit resistance R_t into a water component R_w and an external component R_e. That is

$$R_t = R_w + R_e \qquad\qquad (6.20)$$

By discharging the capacitor first through water and then through a short-circuit across the cup electrodes, the two respective time constants yielded values for the calculation of R_t and R_e. The water resistance was the difference between these two figures.

		(a)	(b)
Capacitance	C	0.5 µF	2.0 µF
Charging voltage	V_o	6.0 kV	3.0 kV
Stored energy	E	9.0 J	9.0 J
Undamped current amplitude	I_o	78.8 A	93.8 A
Discharge time constant	T	0.41 ms	0.28 ms
Total circuit resistance	R_t	4.29 Ω	6.29Ω
External resistance	R_e	1.61 Ω	0.58Ω
Water resistance	R_w	2.68 Ω	5.71 Ω
Energy dissipated in water	E_w	1.70 J	3.51 J

Table 6.1 : Measured and derived quantities for an arc and arcless discharge

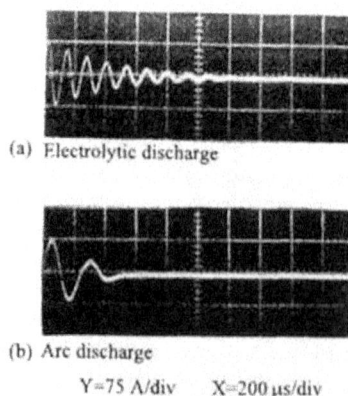

(a) Electrolytic discharge

(b) Arc discharge

Y=75 A/div   X=200 µs/div

Figure 6.3 : Current oscillograms for points (a) and (b) of figure 6.2

The energy dissipated in the water is given by the action integral of Eq.6.14 multiplied by R_w. The result is shown in the last line of table 6.1. It indicates that about twice as much energy is absorbed in the plasma discharge than in the silent electrolytic discharge. Most of the difference, which came to 1.81 J, must have been dissipated in the small arc itself. The arc was observed as a small bright spot on the central electrode, and thus occupied only part of the current path between the rod-and-ring electrodes.

This energy difference could have been consumed in the evaporation of less than 1 mm^3 of water, or it might have been converted directly to mechanical work, via a back-e.m.f. concealed in R_w, for lifting water and the 2.8 gm metal weight on the wooden float of figure 6.1. To decide between these two alternatives, we have to look at the little steam bubble which might have been created by the arc close to the copper stem. This bubble would have exerted the same pressure in all directions on the surrounding water, and this pressure would have been transmitted to all parts of the cup contents. With the float and metal weight removed from the water surface, the explosion should have lifted the water level everywhere in the cup. Instead, as was pointed out before, the arc threw up a narrow water column and drops of water directly above the stem electrode. This reminds one, of course, of the head of the mercury fountain of Chapter 2. The formation of the fountain effect is not compatible with the steam bubble theory. The fact that no water splashed out of the cup when the float and weight were in place provided further evidence of the absence of a steam bubble.

The upward directed action in the saltwater cup experiments hinted strongly at the existence of longitudinal Ampère forces in the plasma discharge. This was the first experimental evidence suggesting that water arc explosions were driven by electrodynamic forces.

Some idea of the order of magnitude of the water lift force can be obtained from point (c) on figure 6.2. The capacitor bank was initially storing 14 J, and the explosion hoisted the 2.8 gm mass approximately one centimeter in the air. With Eq.6.5 it translates to an initial velocity of $v_0 = 0.44$ m/s. Hence the electrodynamic impulse from Eq.6.6 must have been 1.24 mNs. The current oscillogram indicated a time constant of T= 0.23 ms. If this is equated to the pulse duration Δt in Eq.6.7, we find the time average force to be <F>=5.4 N.

With respect to the mercury fountain it was argued that the Lorentz pinch force could not set continuous liquid circulation in motion, and no water circulation appears to take place in the saltwater experiments just described. Therefore, the pinch force could possibly jettison a water column into the air. According to a formula derived by Northrup [2.8], the pinch pressure on the water column above the copper rod could exert an end-thrust on the rod of

$$ F_{th} = \frac{1}{2} \left(\frac{\mu_0}{4 \pi} \right) I_0^2 \qquad (6.21) $$

Comparing this formula with Eq.6.15 indicates that the dimensionless constant k in Northrup's formula is 0.5. In the previous example for point (c) in figure 6.2, I_0 was 104.7 A. Therefore the pinch thrust lifting the water column came to 0.55 mN. This was the maximum -- not the average -- pinch thrust. Thus the average force was at least four orders of magnitude too small to impart the observed momentum to the 2.8 gm mass. Aspden's [6.3] calculations revealed a similar discrepancy. Hence we can rule out Lorentz pinch forces as being the cause of water plasma explosions.

More evidence against thermodynamic arc explosions will be outlined in the remainder of the book. A unique facet of the saltwater cup experiments was the impotence of electrolytic currents to generate any observable electrodynamic forces.

The two examples of table 6.1 deal with this surprising discovery. An electrolytic current pulse of 78.8 A amplitude left the water undisturbed. A 93.8 A amplitude through a plasma caused an explosion. It can only mean that the current elements of a non-electronic current do not interact with each other as strongly as two elements in metal or plasma would do. Furthermore, there appears to have been no interaction between metallic current elements in the copper electrode and electrolytic elements in the saltwater.

Arcs in a Saltwater Cartridge

Frungel [6.6] discovered the working principle of water arc launchers in 1947. His apparatus is shown in figure 6.4. The arc was established between a vertical rod electrode and a coaxial ring electrode. The rod passed through a dielectric plate which supported the ring. Water filled the annular cavity between the electrodes. In most cases this seems to have been tap water. A thin sheet of mica was placed on top of the ring so that its underside touched the water. On the sheet rested a small metal disk projectile.

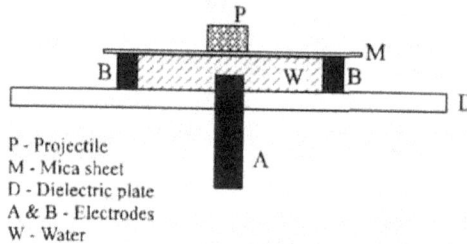

P - Projectile
M - Mica sheet
D - Dielectric plate
A & B - Electrodes
W - Water

Figure 6.4 : Frungel's water-arc launcher

By discharging a 12.2 kV, 70 nF capacitor through the water, Frungel was able to catapult a two-gram mass two meters up in the air. The projectile travelled with an initial velocity of 6.26 m/s and the current pulse lasted about 50 μs. For these figures Eq.6.7 yields an average force of 250 N. The mica sheet stayed in place, proving there was no residual steam pressure present which could lift the sheet after the projectile had departed. Frungel concluded there was no steam present and a satisfactory explanation of the launch force had not been found.

Almost forty years elapsed until the authors' group turned its attention to powerful water plasma explosions [6.7], albeit without any knowledge of Frungel's pioneering work. It was the momentum technique of measuring arc forces which made us think of using water plasmas to launch objects into space. At that time electromagnetic launchers were receiving considerable publicity, because ordinary guns, relying on chemical explosives, had been found incapable of launching projectiles into orbit around the earth.

The effectiveness of all electromagnetic launchers depends on the k-factor of Eqs.6.15 and 6.17. This dimensionless parameter is independent of the size of the launcher and permits small-scale tests in the laboratory. In Frungel's shot the performance index was k = 190. This compares with 5<k<10 for railguns. Already in the mid-1980s it was clear that water plasma launchers would be far more effective than the railgun and other electromagnetic launchers, provided the arc could be contained without destroying the launch device. The saltwater cup was very weak and something stronger had to be found.

The first improvement over the plastic cup was the dielectric cartridge of figure 6.5. Its body was a block of glass-fibre reinforced epoxy, known as G10. Fibre-glass mats made the block a laminated structure. Ultimate failure of the block was caused by splitting the laminations apart. One-half-inch square copper bars were tightly fitted into a milled groove of the dielectric block, and set flush with its upper surface. A ½-inch long gap was left in the center between the copper bars. It formed a ½-inch cubic cavity which was filled with saturated saltwater. Axial motion of the bars was prevented by four horizontal bolts passing through the bars and the dielectric cartridge. A small dielectric plate was placed on top of the saltwater cavity. A metal weight rested on top of the plate. To prevent flashover in a thin layer of air between the water and the dielectric cover plate, a ½-inch square and shallow projection was machined on the underside of the cover. This dipped into the water and displaced the air.

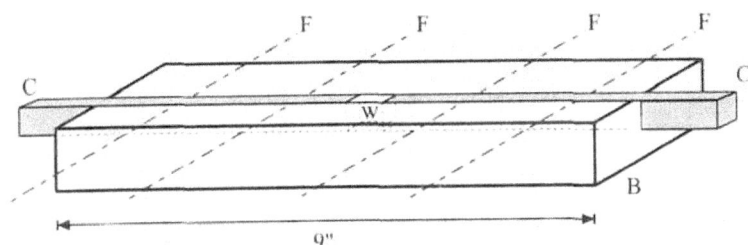

B - Glass-Fibre Epoxy Block
C - 1/2" x 1/2" Copper Bars
W - 1/2" Cubic Water Cavity
F - Restraining Bolts

Figure 6.5 : Dielectric cartridge

It will be realized that the dielectric cartridge is analogous to the straight-through liquid mercury channel of figure 2.11. This had proved how longitudinal Ampère forces make themselves felt in a liquid conductor. It was hoped that water plasma, instead of liquid mercury, would be subject to the same set of forces.

As in the saltwater cup experiments, small values for the product of CV_0, resulted in silent electrolytic discharges through the saltwater. The magnitude of the silent current was stronger in the cartridge than in the cup configuration. Some electrolytic conduction appears to take place at the start of any discharge pulse through water. The electrolytic conduction phase partially discharges the capacitors and, therefore, must subtract from the strength of the subsequent arc explosion. This reduction does not appear to be the same in successive shots and leads to scatter in the force measurements. Nevertheless, this investigation clearly revealed the major characteristics of the water plasma explosions.

A number of the experimental shots damaged the cartridge assembly. At the very highest pressures generated in the explosion cavity, water would be driven between the copper bars and the epoxy. Sometimes this caused the conductor bars to bend upward. More serious was delamination damage of the cartridge body. On one occasion a complete layer of the laminated structure was pushed sideways out of the block. Sooner or later the laminations would part and permit water to leak out of the cavity. This made frequent repair and replacement of the cartridge necessary. Nevertheless, a large number of shots were fired in which most of the explosion energy was imparted to the projectile.

The principal measurements made during every shot were the recording of the current pulse, as in the cup experiments, and the determination of the throw-height h of the projectile of mass m. The measured mechanical impulses exerted on the projectile by four of the most powerful plasma arc explosions are plotted on figure 6.6 against the action integral. Table 6.2 lists the experimental results and calculated quantities of the four shots plotted on figure 6.6.

Shot #	C μF	V_0 kV	I_0 kA	T μs	m kg	h m	$\int F\,dt$ N s	$I_0^2 T/4$ A^2 s	k	cartridge
20	8	15	12.7	65	0.977	0.123	1.52	2621	5799	old
25	8	20	16.9	65	0.977	0.508	3.08	4641	6637	new
26	8	25	21.2	65	0.977	0.950	4.21	7303	5765	new
27	8	30	25.4	65	1.597	0.975	6.98	10484	6658	new

Table 6.2 Dielectric cartridge shots displayed in figure 6.6

Recalling the definition of the action integral A by Eqs.6.12 and 6.14, the straight line on figure 6.6 is represented by

$$ m\,v_0 \; = \; \int_0^\infty F\,dt \; = \; \left(\frac{\mu_0}{4\pi} \right) k\,A \qquad\qquad (6.22) $$

where $A = (I_0^2 T/4)$. Hence k can be derived from the slope of the straight line on figure 6.6.

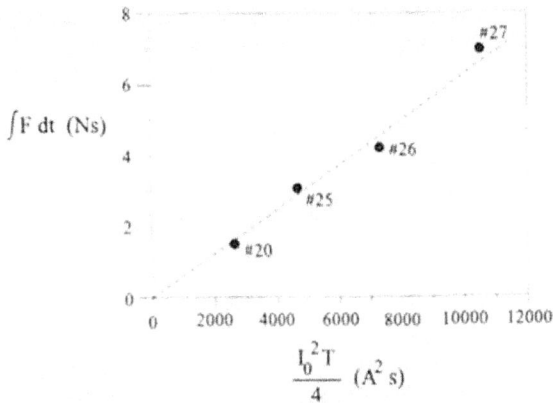

Figure 6.6 : Dielectric cartridge results

The four plotted shots, involving impulses of 1.5 to 7.0 N s (or the corresponding initial momenta in kg m/s) then yield k=6700. This is a very high k value for a pulse-type electrodynamic accelerator. It compares with k~10 for railguns and k~200 for one of Frungel's water-arc explosions. The cartridge results, more than anything else, encouraged the further study of water arc launchers.

If the object is to launch projectiles into space, does it matter if the water arc explosion is driven by thermodynamic or electrodynamic forces? It certainly does. As already mentioned, chemical explosions, based on the expansion of hot gases, have not been able to propel objects into earth orbit. The inherent velocity limitation of any adiabatic gas expansion would also apply to superheated steam. The designers of space catapults must therefore pin their hopes on electrodynamic forces, such as those indicated by the straight line of figure 6.6. Thermodynamic forces are unlikely to be governed by a similar linear relationship. Even though the amount of Joule heat generated in an electric conductor is proportional to the action integral, the latent heat of evaporation would make the relationship between force and action integral non-linear.

To obtain an idea of the amount of energy which was supplied to the water arc during the most powerful shot #27, the cartridge was replaced by a solid copper bar of the same length and cross-section. When discharging 8 μF at 30 kV through this conductor, the short-circuit time constant was found to be T_{SC} = 255 μs. If R_{SC} is the short circuit resistance of the circuit, then using Eq.6.11 this may be equated to

$$R_{SC} = \frac{2L}{T_{SC}} \qquad (6.23)$$

The selfinductance L depends only on the geometry of the discharge circuit. This was not changed and remained $L = 11.1$ μH. In this way it was found that $R_{SC} = 87$ $m\Omega$. If an effective resistance R_a is assigned to the water arc, which must allow for the induced back-e.m.f., then the total resistance in shot #27 was R_a+R_{sc}. With a time constant of 65 μs for the water shot (see table 6.2), the combined resistance comes to 342 $m\Omega$. By calculating the difference, the effective water resistance was found to be 255 $m\Omega$. The total energy stored in the capacitor bank before the shot was

$$E = \frac{1}{2} C V_0^2$$ (6.24)

where $C = 8$ μF, $V_0 = 30$ kV and therefore, E=3600 J. The fraction $R_a / (R_a+R_{SC})$ of E was dissipated in the water and came to 2684 J, which is equal to 642 cal.

The plasma cavity contained 2 cm^3 of water with a latent heat of evaporation of 529 cal/cm^3. Hence the energy deposited in the water was insufficient to evaporate it all, let alone raise the steam to the explosion pressure. That all the water in the cavity formed part of the exploding plasma was deduced from photographic studies. Open-shutter photographs of a typical explosion when the cavity was open and the plasma was allowed to expand freely out of it, as in figure 6.7, display an external plasma cloud that was many times the volume of the explosion cavity. There is not the slightest doubt that every bit of the water was ionized and contributed to the explosion force.

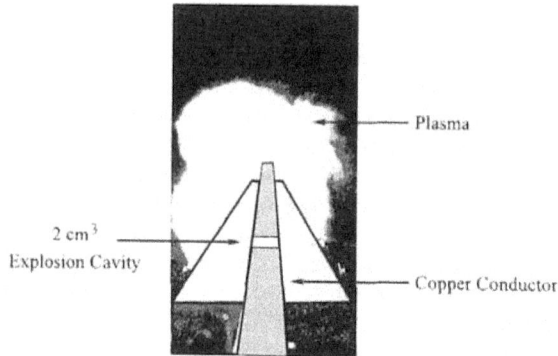

Figure 6.7   Photograph of plasma ejection from the open cavity of the saltwater dielectric cartridge

Dragone [6.8] had previously investigated the small steam bubble hypothesis. He analyzed the steam temperatures and bubble volumes which would have had to exist to explain the measured explosion forces, and arrived at temperatures as high as 55 million degrees Kelvin in volumes as small as 10^{-4} cm^3. These calculations show that unrealistic temperatures and volumes are required if a thermal process is involved. In fact, if such temperatures existed somewhere in the arc, a thermonuclear reaction could occur. For this reason we feel that the mechanism involved in water arc explosions is not fully understood.

Another mechanism must be sought to explain water arc explosions. Perhaps it is safer to explain the arc explosions phenomenologically as an electromechanical conversion process.

In yet another experiment, a weight of several kilograms was placed on the explosion cavity of figure 6.5. When 3600 J were discharged through the arc, and the plasma would have completely filled the cavity, the weight barely lifted, allowing only a drop of water to escape. When the weight was taken off the cavity, the water was found to be hot but not boiling. Any thermal explanation of the explosion which is based on extensive water evaporation and steam superheating is therefore not viable.

At this stage of the investigation attention was directed to reinforcing the explosion cavity so that more powerful current pulses could be pushed through the saltwater. The design effort quickly led to traditional gun configurations which contain the explosion in a substantial steel barrel.

Miniature Water-Arc Gun

None of the mechanisms of heat generation, and particularly not ohmic heating by electron impact, depend strongly on the geometry of the current path through the water, so long as the current density and the water volume remain the same. Changing the electrode geometry from the cartridge to a gun configuration should, therefore, generate approximately the same water heating, for equal current pulses and arc volumes. Tests soon revealed, however, that the gun geometry greatly reduced the intrinsic strength (k value) of the explosion. This can only be understood on the basis of electrodynamic forces which are strongly influenced by the current path geometry.

Figure 6.8 is a diagram of a small water-arc gun which was built and tested in the MIT electrodynamics laboratory. The 12.7 mm bore steel barrel was 10 cm long from breech to muzzle. The center electrode consisted of a 6.35 mm diameter copper rod insulated with a 3.17 mm thick nylon sleeve. A fibreglass epoxy block provided secondary insulation. The capacitor discharge current was passed from the copper baseplate through the rod electrode to the saltwater cavity and hence to the steel barrel and out of the heavy current lug.

The performance of this small water arc gun was reported at the 4th Symposium on Electromagnetic Launch Technology held at the University of Texas in April 1988 [6.9]. Table 6.3 lists the results of the first series of measurements. An 8 cm long nylon cylinder of 12.7 mm diameter and weighing 8 gm, formed the projectile. It was placed on a 2 cm long saltwater column. In every shot a fraction of the water was expelled from the barrel and presumably travelled with the projectile. The amount of water which remained behind in the barrel was not measured. For the purpose of the calculations presented in table 6.3, one gram out of 2.5 gm of saltwater was assumed to have travelled with the projectile at the muzzle velocity v_m. Hence the total projectile mass was taken to be 9 gm.

It was planned to measure the projectile velocity with a two-screen-method. The screens were to be mounted a fixed distance apart, and the flight time of the projectile between the screens was to be recorded. While the instrumentation for this was being assembled, tests were started with a single aluminum foil screen. Breakage of the screen by the projectile produced a pulse which was recorded on an oscilloscope. The time base of the scope was triggered by the onset of the capacitor discharge. The miniature gun was damaged before the two-screen velocity measurement became available. Therefore the muzzle velocity, v_m, with which the projectile left the gun, was estimated from the one-screen measurement by the following method.

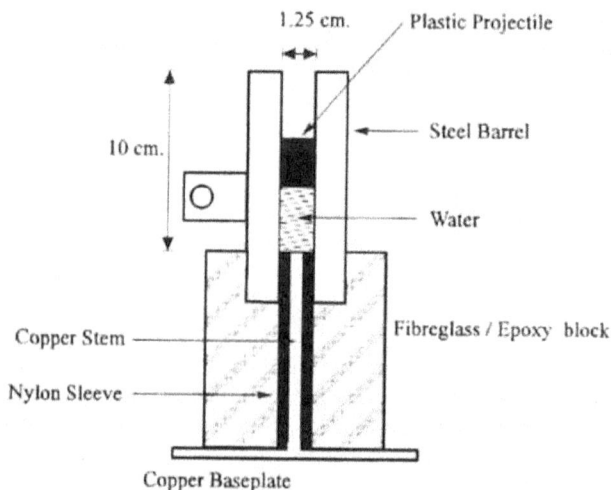

Figure 6.8 : Miniature coaxial water-arc gun

Shot #	V_0 kV	Δt ms	v_{av} m/s	v_{muzzle} m/s	efficiency
1	20	1.6	166	182	9.3 %
2	30	1.2	221	243	7.4 %
3	40	1.1	241	265	4.9 %
4	50	0.92	288	317	4.5 %

Table 6.3 . Velocity measurements with 9 gm projectile using the water-arc gun shown in figure 6.8

If s is the distance from the top of the stationary projectile in the barrel to the aluminum foil screen, and Δt is the time required for the projectile to reach the screen, then an average velocity may be defined by

$$v_{av} = \frac{s}{\Delta t}$$

(6.25)

This is less than the actual muzzle velocity because of the acceleration from zero velocity. Conservatively estimated we may assume

$$v_m = 1.1 \ v_{av} \qquad (6.26)$$

Four shots were fired with the 8 μF capacitor bank charged to various voltages up to 50 kV. The current decay time constant for this series of shots was T=160 μs, and the distance to the foil screen was s = 26.5 cm. As shown in table 6.3, the muzzle velocities varied between 182 and 317 m/s.

The water column length probably expanded up to two or three times its original length of 2 cm during the 160 μs of current flow. This meant that the current streamline pattern in the water was continuously changing, thus giving a varying instantaneous value of the dimensionless constant, k, during the shot. This seems to be reflected in the fact that the highest energy shot yields the lowest efficiency.

In the second series of experiments the nylon projectile was shortened to 1 cm, and it was placed on top of a 1 cm long saltwater column. The total mass (water + projectile) expelled from the gun barrel at the muzzle velocity was assumed to have been 1.5 gm. In these measurements the foil screen was 32.3 cm above the top of the stationary projectile. The time constant was found to be 110 μs. All results obtained with the small nylon projectile are listed in table 6.4. As expected, for the same capacitor voltage, the smaller projectile was accelerated to a higher velocity. In the four shots up to 50 kV, the muzzle velocity increased to 770 m/s.

Shot #	V_0 kV	Δt ms	v_{av} m/s	v_{muzzle} m/s	efficiency
1	20	0.73	441	485	11.0 %
2	30	0.58	555	611	7.8 %
3	40	0.51	631	695	5.7 %
4	50	0.46	700	770	4.4 %

Table 6.4 Velocity measurements with 1.5 gm projectile using the water-arc gun shown in figure 6.8

At the end of the second series of experiments the breech of the gun began to disintegrate. Holes in the nylon insulation were discovered by probing with a thin rod. A few more shots were attempted with just saltwater and no solid projectile. It was of interest to find out how effective a projectile liquid water would be. This test produced the most dramatic result. A 3.8 gm mass of ionized water was ejected from the barrel at an estimated velocity of 1000 m/s. A ¼-inch thick aluminum barrier plate had been set up 10 cm above the muzzle. On top of the plate, and coaxial with the gun barrel, was a die of the same diameter as the bore of the gun. The ionized water projectile flew through the atmospheric air and then crashed through the aluminum and punched a 12.7 mm hole through the target plate. The water was captured on the other side of the plate. It was found to be lukewarm and still contained all the salt in solution. This result could certainly not have been achieved with superheated steam. It also proved that, at high velocity, the inertia of the water molecules prevented significant lateral deflection of the liquid and the water projectile appeared to be as effective as a solid object of the same mass would have been. Figure 6.9 shows the hole which the water punched through the aluminum plate.

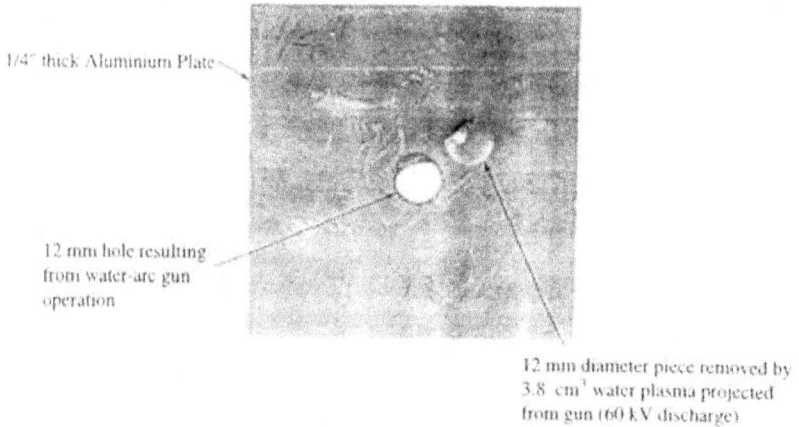

1/4" thick Aluminium Plate

12 mm hole resulting from water-arc gun operation

12 mm diameter piece removed by 3.8 cm^3 water plasma projected from gun (60 kV discharge)

Figure 6.9 : Puncture of a 1/4" aluminum plate by a slug of water.

The development of electromagnetic launchers aims to achieve the highest possible projectile velocity combined with maximum kinetic energy. By computing the efficiency of energy conversion for the recorded shots of tables 6.3 and 6.4, it is found that the highest efficiency of 11 percent was obtained with the smaller mass (1.5 gm) and the least stored energy (1.6 kJ at 20 kV). The experiments show a decrease in efficiency with increasing projectile mass and energy stored in the capacitors. It is as if the inertial resistance of the projectile forces the input energy to flow into an alternative channel, other than kinetic energy.

Unfortunately, the results of tables 6.3 and 6.4 are not strictly comparable because of differences in water volume. The variation of k in Eq.6.22 with water expansion would have been different in the two instances, and thus with an identical current pulse, different impulses ($\int F\, dt$) would have been applied to the two masses. The expansion of the water volume and consequent time variation of k were avoided in experiments with a very large lifting mass, such as those listed in table 6.2. Unfortunately, in these shots there were no experiments with different masses, accelerated by the same initial capacitor voltage.

In any experiment in which F(t) is the same for two explosions propelling different masses m_1 and m_2 ($m_1 > m_2$), the energy diverted to straining the cavity walls and breaking atomic bonds should not vary. According to the laws of conservation of momentum and energy, if these two objects have equal momentum as a result of equal applied impulses, then the less massive one, m_2 acquires more kinetic energy than m_1. One might then wonder where the additional kinetic energy of the lighter mass comes from, since both shots were supplied with the same initial energy, $\frac{1}{2}CV_0^2$.

Since the two current pulses, as well as the water volumes, are assumed to be identical, there should be no difference in the Joule heat generated in the plasma by electron-ion collisions. Thus the only way of explaining the difference in kinetic energies appears to be that the non-thermal forces initially accelerate ions in specific directions. However, on collision with massive objects, such as the walls or the projectile, these ions become thermalised and thus get hotter. In this way the kinetic energy difference between m_1 and m_2 is easily explained. The inertia of the greater mass forces more of the mechanical energy into heat. By the same token, maximum kinetic energy is acquired by the smallest possible mass.

There are, however, practical limits to the kinetic energy that a small mass may acquire. Its rapid acceleration will produce changes in the current streamline pattern in the water. This is likely to reduce k in Eq.6.22, and with it the force $F(t)$ which is ultimately responsible for the projectile velocity.

The most puzzling aspect of the water plasma explosions is the experimental revelation of very high values of the dimensionless performance coefficient, k in Eq.6.22, which in some cases exceeded 6000. It seems virtually impossible that any electrodynamic force law alone could fully account for the exceptionally large explosion forces.

Due to the large k-values, it is not at all certain that the water plasma launcher can be scaled up according to the laws of Newtonian electrodynamics, however there is little doubt that it could outperform the railgun and other electromagnetic launchers. Water plasma propulsion probably presents the best chance of catapulting small masses into space without the use of rockets.

Air-Arc Explosions

Pulsed arcs in air can be formed without an arc cavity. Experiments of this type will be referred to as 'open arc explosions'. An enclosure is required if the arc pressure is to be measured by the momentum technique which was originally developed for water-arc explosions. In the latter case we will speak of 'entrapped arcs'. The air pressures developed inside and outside the arc differ greatly for open and entrapped arcs.

It was found to be easy to use the dielectric cartridge of figure 6.5 for the entrapped experiments. The procedure was the same as that for water arcs, except that the cavity contained laboratory air at atmospheric pressure and temperature.

Using the same notation as for the water arcs, the results of five entrapped air-arcs are listed in table 6.5. Measured mechanical impulses, imparted to metal projectiles, ranged up to 1.32 N s. With a one-gram projectile this had the potential of producing an initial velocity of 1.32 km/s. Sufficiently large current pulses through atmospheric air-arcs thus appear to be able to compete with ordinary guns relying on chemical explosives. A loud noise, similar to gun fire, is also heard when air-arcs explode. It is so loud that experimenters have to wear protective ear-defenders.

Shot #	C μF	V_0 kV	I_0 kA	T μs	m kg	h cm	$\int F\,dt$ N s	$I_0^2 T/4$ A² s	k
1	8	20	19.5	200	0.3706	3.6	0.31	19,013	163
2	8	25	24.0	220	0.3706	4.9	0.36	31,680	114
3	8	30	27.0	225	0.580	11.2	0.86	41,006	210
4	8	35	33.0	240	0.580	26.5	1.32	65,340	202
5	8	40	38.3	225	0.580	24.5	1.27	82,512	154

Table 6.5 . Entrapped air-arc results

Figure 6.10 is a graph of the measured entrapped air-arc impulses plotted against the action integrals of the current pulses. According to Eq.6.22, the graph should be a straight line. There is no doubt that the explosion force increased with the action integral, but it would be premature to claim strict proportionality. The scatter of the experimental points on figure 6.10 may be due to some non-systematic disturbance. For example, varying amounts of surface discharge current flowing over the cavity walls could have subtracted from the strength of the explosions. Direct proportionality, as represented by the straight line of figure 6.10, however cannot be ruled out.

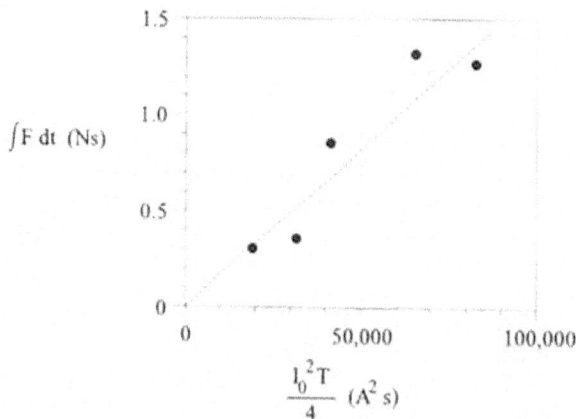

Figure 6.10 : Projectile momentum as a function of the action integral through air in the dielectric cartridge shown in figure 6.5

When comparing the air-arcs with the water-arcs of figure 6.6, a most significant difference emerges. For saltwater the dimensionless performance coefficient k had a value of 6700, while in the entrapped air-arcs it fell to 170. This forty-fold decrease in the effectiveness of the explosion is difficult to explain. At first sight, none of the electrodynamic theories predict that the force is a function of the conductor material, but we now find that electrodynamically driven arc plasma explosions are very much dependent on the arc medium. There is, however, the consideration that in the metal lattice the magnetic force on an atom is passed on to the lattice collectively. The situation is different in a liquid conductor, where the force will be mechanically transferred to adjacent atoms that lie along the direction in which the force acts. In this way the longitudinal Ampère force accumulates along a chain of atoms lying on a current streamline. Hence we have reason to expect higher k-values in liquid conductors than in solids. This aspect will be analyzed in the next chapter in connection with capillary fusion.

Computations with the Ampère force law have shown that in wire circuits k - 10 and at most 20. The contact magnification of force in liquid conductors can raise the k-value as high as 100, and the air-arc result of 170 does not seem impossible. The step-up of k in water to 6000 is grossly out of line with any contact magnification process. It probably implies a new and unknown force generating mechanism.

One is tempted to think that the difference between water and air is due to the compressibility of air and the fact that the current elements in air are not in contact with each other. 'Being in contact', however, has no precise meaning. On the average, an atmospheric air molecule incurs about 7000 collisions per microsecond. Since a typical capacitor discharge pulse lasted for about 100 μs, the air must have looked almost like a liquid. In any case, k = 170 would be difficult to justify if the air plasma was a collection of isolated ions. In the event we might argue that the air plasma behaves like a normal plasma conductor, while water shows abnormal behavior.

Let us now consider open air-arcs and particularly the question of whether these plasma explosions are driven by thermodynamic or electrodynamic forces. There should be a decisive difference in the collective particle motion between an adiabatic gas expansion and an electrodynamic plasma explosion. The thermal expansion should rely on randomized particle collisions and the consequent isotropic advance of the shock front. On the other hand, electrodynamic forces should be more organized in certain directions and are unlikely to form a spherical shock front around a short arc. If they are to produce a large outward radial pressure, the electrodynamic forces must, either, act radially outward, or they must be axial expansion forces which, because of the instability of the plasma column, break out in radial motion. Positive ions will take along electrons as in ambipolar diffusion [6.2]. Once ejected from the arc gap, the ion pairs find themselves in a largely field-free region and should obey ordinary gas dynamics. Electrodynamically accelerated ion pairs, with entrained neutral particles, if moving at supersonic velocity should give rise to a wedge-shaped shock front in the ambient air. Since the plasma particles continue to emit light after they have been ejected from the arc region, it should be possible to photograph the shape of the shock front. A spherical shockwave would confirm thermal expansion and a tapering wedge-shaped shock front would prove supersonic radial expansion, driven by electrodynamic forces.

With this in mind, experiments were performed with open air-arcs across a 3.2 mm gap between the ends of 6.35 mm diameter stainless steel rods. An 8 μF capacitor bank was discharged through the gap in accordance with Eq.6.8. In successive shots I_0 varied between 20 and 65 kA. Even though the arcs were quite short, they appeared as blinding flashes illuminating the darkness of the laboratory. Radiated energy from the arcs excites the emission of ultraviolet light in the surrounding air, which looks like a ball of intense light. Trial and error experimentation with UV filters, polarizing and different color filters, finally produced open-shutter photographs which clearly revealed the structure of the arc and the ejected plasma.

Figure 6.11 is an open-shutter arc photograph obtained from a distance of 30 cm and with the camera set to f16, with all the filters in place. The shutter was opened remotely, in the dark laboratory, before the capacitor bank was discharged. The arc current pulse appertaining to figure 6.11 was $I_0 = 45$ kA, T = 188 μs. Figure 6.12 has been drawn to explain the features of figure 6.11. The arc region, in which ions and free electrons are created and accelerated toward the metal electrodes E, is denoted by A. The strong electric field is confined to this arc region. Surface melting occurred where the arc was rooted on the metal electrodes and where much of the ion recombination heat must have been liberated. After ten shots the outward flow of molten metal had made the electrode ends dome-shaped. Pronounced lips had formed at the edges, as shown in figure 6.12. The arc enveloped the lips and thereby concealed their outline on the photograph. On color photographs the region A was bright amber. Lateral jets J can be seen to emerge from the arc region. The jets focused on an edge which, when photographed from above, was seen to be a circle. In three dimensions the J region had the shape of a discus. Its color was also bright amber, fading toward the edge. A sharp boundary can be seen to separate the J region from the surrounding plume P. The latter was dark red, distinctly different from the amber of the jets. In successive shots, with the same current pulse, the J region was always the same shape and size. The outline of the plume varied, however, from one shot to the next, presumably on account of air drafts.

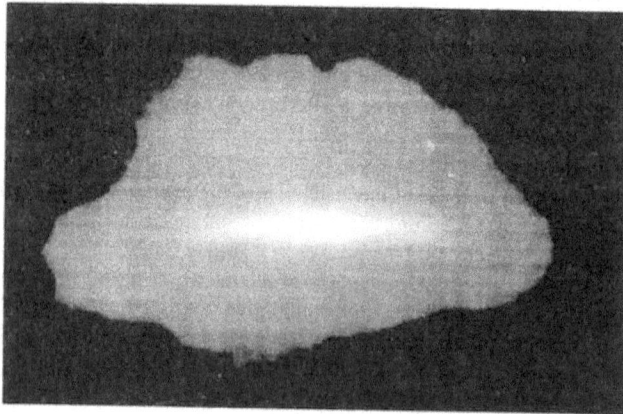

Figure 6.11 : Open shutter photograph of an atmospheric air arc

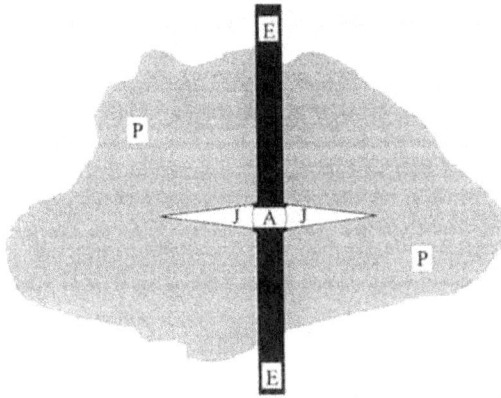

Figure 6.12 : Arc and shock-front structure

The outline of the J region appears to be the supersonic shock front punching into still air, at the time the light emission became faint. The wedge-shaped plasma stream confirms the radial ejection of ions from the arc. The shock front is certainly not of the spherical shape that would be produced by thermal plasma expansion. The wedge is due to the ablation of ions which scatter into the ambient air. The ablated and scattered ions then undergo many collisions with air molecules and drift with the body of the surrounding air. This explains the existence of the red plume. Photographs like figure 6.11 furnish decisive evidence that arc expansion is not dominated by a thermal mechanism.

The Cause of Thunder and Retrograde Lightning

Thinking man has observed thunder and lightning for more than a million years, and this natural spectacle plays a powerful role in both eastern and western mythology. Thunder is still recognized as one of the oldest riddles in recorded scientific speculation. At the end of the twentieth century we are still arguing about its cause. Remillard [6.10] published an excellent review of thunder research up to 1960. He quotes Aristotle's opinion of thunder being caused by the sudden ejection of vapors from the clouds. The vapors subsequently caught fire. Two hundred years later Lucretius was the first to attribute thunder to the collision of clouds. His view appears to have been shared by Benjamin Franklin, nearly 2000 years later. A little earlier the imaginative French thinker René Descartes modified the Lucretius explanation by suggesting that one cloud falling on another caused the rumble in the sky. In all these years lightning was never considered to be the cause of thunder.

Today we have no doubt that without lightning there would be no thunder. The noise from remote parts of the lightning stroke reaches us later than the sound from nearby sections. With sound travelling at 330 m/s, and thunder being audible up to distances of the order of 15 km, the roll of a single stroke may last the best part of a minute.

In contrast to rolling thunder, a ground stroke a few meters away from an observer sounds like a powerful crash, not unlike the explosion of a large quantity of dynamite. What kind of force sets the air molecules in motion and produces the shockwave in the atmosphere?

The modern conventional answer seems so simple. Lightning creates a crooked plasma channel, composed of white hot ionized air. It is heated so quickly, in a fraction of a millisecond, that its lateral thermal expansion is not unlike the explosion of gun powder. Travelling away from the lightning channel, the supersonic shockwave degenerates into an acoustic disturbance proceeding at the reduced speed of 330 m/s. This persuasive argument of the past fifty years smothered further inquiry into alternative causes of thunder. It has been difficult to substantiate the thermal shock explanation with factual evidence, but no other theory was put forward.

Then, in 1961, Viemeister [6.11] wrote:

> "Cold lightning is a lightning flash whose main return stroke is of intense current but of short duration. Hot lightning involves lesser currents but of longer duration. Hot lightning is apt to start fires while cold lightning generally has mechanical or explosive effects."

Although Viemeister's statement was in stark conflict with the prevailing thunder theory, he provided no argument of how cold lightning could produce even louder thunder, and therefore greater pressure, than hot lightning. It was the discovery that arc explosions were far colder than generally assumed, which directed Peter Graneau's [6.12] attention to Viemeister's assertion about hot and cold lightning.

According to Remillard [6.10], thunder research reached a peak in the period 1870-1920, which coincided with the evolution of modern electromagnetic field theory. By that time thunder was clearly recognized to be the result of lightning, and the acoustic effect had to be spread out all along the lightning channel. Four theories were competing with each other, and the debate between their exponents was published in the Scientific American. One of the theories claimed that lightning was evacuating its channel and thunder was caused by the subsequent implosion of air. This did not survive because one atmosphere of pressure was far too little to account for the strength of the lightning-arc explosion. Two other theories relied on the pressure of water vapor in the air. One claimed that the water molecules were dissociated into hydrogen and oxygen and later recombined explosively. The other was based on steam pressure causing the explosion of the lightning channel. Both theories had to be abandoned when it became clear, from laboratory experiments, that the strong radial expansion of arc plasmas also occurred in completely dry air.

The fourth theory of that period was the thermal expansion model. First proposed by Hirn [6.13] in 1888, it remained undisputed up to the publication of Viemeister's book in 1961. Surprisingly, electrodynamic forces were not mentioned during the hectic period of thunder research. At the end of the nineteenth century, the Ampère-Neumann electrodynamics, which had been prevalent in France and Germany for eighty years, was being superseded by relativistic electromagnetism. The only important electrodynamic force of field theory was the pinch force, and this acted in the wrong direction to explain thunder.

Benjamin Franklin claimed in 1749 that the fire of electricity and that of lightning were the same. He started the experimental investigation of thunder and lightning. The subsequent 240 years of measurement and scientific observation have revealed much about the nature of lightning and the accompanying acoustic phenomena. The difficulty of locating instruments close to the unpredictable stroke channels have left the cause of thunder to speculation rather than hard experimental evidence.

The many observations made since Franklin's time cover the measurement of electric and magnetic field strength in the vicinity of ground strokes and at high altitude, the recording of lightning current pulses, the analysis of the sound spectrum resulting from thunder, and the analysis of the electromagnetic radiation spectrum emitted by lightning. Most relevant to the cause of thunder should have been atmospheric overpressure measurements, however, they had to be made some distance away from the channel of naturally occurring lightning discharges, and are therefore an unreliable guide to shock front pressure.

Very discriminating observations regarding the sounds of thunder have been made by the human ear. All descriptions agree on the existence of two phases of thunder consisting of the initial crash, as Lucretius called it, and the subsequent long roll. The crash seems to be loudest close to the point where lightning strikes the ground. It has been claimed to be the most intense of all noises produced by nature. It instills fear in human beings and in animals. That the crash, also known as the peal or clap, may be just as intense higher up in the atmosphere is eminently reasonable. The proof of this has still to be provided.

The pressure in the shock front of lightning sets air molecules in motion. Interchanges of kinetic and pressure energy in the air generate sound vibrations which propagate through the atmosphere. Shape and amplitude of lightning current pulses vary. This should give rise to similar variations in the acoustic phenomena. Several investigators have measured and analyzed the sound spectrum of thunder [6.14, 6.15, 6.16].

An account of thunder pressure measurements will be found in Uman's book [6.17]. The major uncertainty was the distance from the lightning stroke at which pressure recordings were obtained. The figures are not a reliable guide to shock front pressures.

The infrared, visible, and ultraviolet spectrum of lightning flashes has been extensively researched over a period of more than one hundred years [6.17]. With regard to the cause of thunder, the most interesting aspect of the lightning spectrum is the information it yields on channel temperature.

Spectroscopically determined lightning temperatures range from 5000 to 36,000 °K [6.18]. They have not been correlated with lightning current which may vary between 20,000 and 200,000 A. It is generally agreed that Joule heating is the primary means of raising the plasma temperature. This heating depends on the square of the current. The 7:1 temperature range so far reported seems very narrow when compared with the 100:1 ratio of the square of the current.

According to Dragone [6.8], the low measured lightning temperatures would be quite insufficient to produce the force of thunder. What has been important in arc and lightning research is the action integral, Eq.6.12, of the current pulse. It has become customary to compare lightning strokes on the basis of their action integrals. Berger [6.19] pointed out that this integral is responsible for the thermal and mechanical effects. The severity of thunder increases with the action integral. Berger mentioned the observation of giant flashes with action integrals up to 2.2×10^7 A^{2}s. These powerful strokes did not kindle forest fires.

Viemeister [6.11] described them as cold lightning.

If the plasma pressure is due to the entrapped arc temperature, it should approximately obey the ideal gas law for constant volume

$$\frac{T_2}{T_1} = \frac{P_2}{P_1} \qquad\qquad (6.27)$$

where the subscript 2 refers to the final and 1 to the initial conditions. Shot #5 from table 6.5, revealed a force which was equivalent to an average pressure, P_2, of 409 atm. Equation 6.27 then calls for an average temperature, T_2, of 120,000 $°K$. This is more than three times the highest lightning temperature ever observed. This very high temperature should have been generated by the relatively small $I_0 = 38.8$ kA, while lightning currents up to 200 kA have been measured.

Against this it may be argued that the cartridge measurements apply to entrapped arcs whereas lightning explosions are open. Pressures and temperatures of open arcs will not reach the values of entrapped arcs. But the thermal theory of thunder [6.16, 6.17] is based on the existence of shockwaves which imply that atmospheric pressure and temperature just ahead of the shock are those of ambient air. In other words the shock front is assumed to entrap the plasma.

Many aspects of our understanding of thunder are undisputed. Everyone agrees that the loudness of thunder is a function of the stroke current. It is not the amplitude of the current pulse, but the action integral of the pulse, which controls the shock and acoustic phenomena. Thunder is certainly due to a shockwave of arc plasma and entrained air emanating from, and all along, the lightning channel. The outward travelling shockwave later degenerates into a sound wave. Transport delays of sound are responsible for the roll of thunder. None of these agreed facts touch the root cause of thunder, that is: what kind of force drives the shockwave?

Mathematically based thunder theories [6.16, 6.17] have been developed in the twentieth century, at a time when field theory was the only model by which electric and magnetic actions were allowed to be explained. The ponderomotive force of field theory is the Lorentz force. Its principal effect is to urge the current carriers in the lightning discharge to be pinched inward into the channel. The pinch force, apart from being quite small compared to the power of thunder, acts in precisely the opposite direction to the forces which drive the atmospheric shockwave. Faced with this situation, investigators had little choice but to turn their attention to the thermodynamics of thunder, and in particular to the thermal expansion of the lightning plasma.

Entrapped air-arc experiments have provided a range of observations that contradict the thermal thunder theory with 'laboratory lightning' phenomena. Direct evidence against the thermal expansion mechanism was provided by photographs like figure 6.11, which prove the organized jet expulsion of plasma rather than randomized thermal expulsion which should have resulted in a spherical shock front.

The evidence for any particular electrodynamic explosion mechanism is not as strong as that against the thermal thunder theory. Since no third alternative has been suggested, efforts are likely to continue to improve the electrodynamic model until it agrees not only qualitatively but also quantitatively with all experimental findings.

The longitudinal Ampère forces in the lightning channel, which seem to be responsible for thunder, produce axial compression of the ion population. The same ions are the Amperian current elements. If the ions had no inertial mass, the pressure would be relieved instantly by longitudinal expansion of the channel. The inertial mass impeding longitudinal motion is not just the mass of individual ions. Rather, it is the mass of substantial air columns which resist the repulsion of neighboring current elements. More than anything else, this accounts for the great pressure in the lightning channel. The pressure can most easily be relieved by radial expansion of the plasma against the pressure of the surrounding atmospheric air. This is what gives rise to thunder.

What happens at the ends of the lightning arc on the surface of the earth and also somewhere in the thunder cloud? The axial pressure will certainly bear down on the earth. The continuation of the current in the ground may, electrodynamically, counteract the pressure force by longitudinal reaction forces. As far as we know, there exists little evidence for a longitudinal explosion of lightning at the point of contact with the ground.

The circumstances are different at the other end of the lightning channel, up in the cloud. We are led to believe that charge flow in the cloud is essentially horizontal and radially inward to the upper extreme of the lightning stroke. At the cloud end of lightning, the axial Ampère pressure can be relieved by shooting plasma, and anything in front of it, vertically upward in the cloud. Because of the density of the moisture, which must consist of tiny water droplets, it is not possible to see the lightning channel in the cloud and any possible upward ejection of plasma.

If, however, the lightning arc terminates near the upper surface of the thunder cloud, it should become possible to observe a luminous fountain of plasma ions thrown up into space. Precisely this phenomenon has been observed by pilots in high flying aircraft, astronauts in returning space missions, and also by ground based observers in the mountains [6.20, 6.21]. These colorful bursts of radiation, which extend up to 90 km above the thunder clouds, have been called "sprites". However, until it becomes clear whether they represent the same phenomena which is predicted by the Ampère forces, we will refer to these events as 'retrograde lightning'. To gain a better understanding of retrograde lightning, the action of Ampère forces at the corner of a circuit has to be examined.

The diagram of figure 6.13 has been drawn for this purpose. It shows three wire branches A, B and C, which are insulated and fixed to the laboratory frame. Between them is a volume of plasma which allows the vertical current i in A to be split into two i/2 branches. In particular, the diagram delineates portions (a), (b) and (c) of Amperian current filaments in the body of the plasma. For simplicity, all the other current filaments in the plasma will be ignored.

Filament (a) carries the full current, i, and (b) and (c) each carry half of the vertical current. We now consider the electrodynamic forces, according to Ampère's law, which current elements in A, B and C, exert on current elements in (a), (b) and (c). First we look at the A-a interactions. They push strongly upward on (a). The great strength of this force is due to the fact that both A and (a) carry the full current, i, and that they are neighboring sections.

The action of B on (a) is a much weaker repulsion because the current in B is only i/2 and the distances between elements are greater. The B-a repulsion has a downward component which subtracts a little from the upward A-a force. So does the C-a repulsion. The horizontal component of the B-a repulsion cancels the symmetrically disposed, and oppositely directed, horizontal component of the C-a repulsion. The dominant force on the corner plasma, therefore, is a strong upward push.

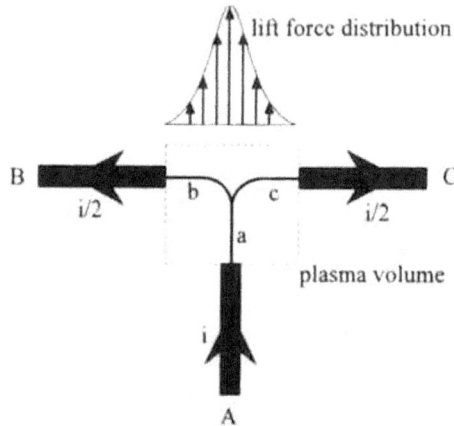

Figure 6.13 : Ampère forces on a plasma current filament at a circuit branch point

The A-b and A-c interactions do cause some lift in the plasma, but it is weak since (b) and (c) carry only half the current and are not located close to A. The horizontal components of A-b and A-c will cause radial expansion of the plasma. Now the B-b, C-c, B-c and C-b interactions are primarily opposing horizontal forces, however, they also create small vertically downward directed forces, on the corner portions of (b) and (c) which subtract from the general upward thrust on the plasma.

Hence the net effect of the electrodynamic forces on the plasma volume is expected to be a lift force which is concentrated along the axis of A and generally distributed as drawn on figure 6.13. For current pulses of lightning strength, lasting for small fractions of one second, the result should be an upward directed explosion that shoots plasma vertically up at great velocities.

Interactions between the plasma filaments (a), (b) and (c) do not react against the laboratory frame and are therefore unable to move the center of mass of the plasma volume upward. Thus we must consider what takes the place of the three wire sections A, B and C in an actual thunder cloud. If B and C are replaced by conducting plasma filaments, which better model the cloud discharge than wires, they will yield to the upward repulsion by A and thereby move upwards, although nowhere near as fast as the ions in (a). Conductor A can be replaced by the lightning column because this column is anchored to the ground by the internal longitudinal Ampère repulsion and the momentary containment of this repulsion by ambient air pressure. In this way the Newtonian electrodynamics explains the phenomenon of retrograde lightning.

An experiment was performed in the laboratory of Hathaway Consulting Services, Toronto, Canada, which demonstrated the existence of a vertical ion jet above a vertical air arc at the corner of a capacitor discharge circuit. The details of the arc gap are shown in figure 6.14. The arc burned between the end-face of a 6.35 mm diameter stainless steel rod and a stainless steel ring of the same outside diameter. The latter was set in an aluminum frame. The ring had a generous aperture for arc ions to escape upward.

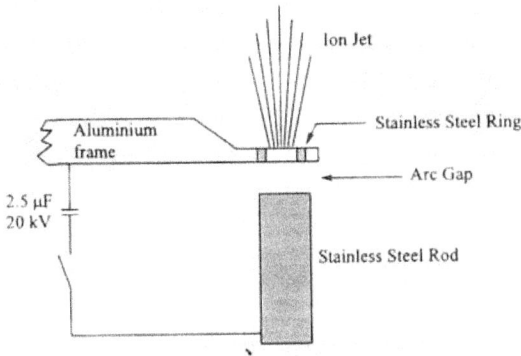

Figure 6.14 : Details of the 22 kA arc gap at the corner of a metallic circuit

The aluminum frame was also struck by the arc and produced aluminum sparks, as in arc welding, which flew upward through the ring and sideways out of the arc gap. The aluminum sparks can be seen in the photograph of figure 6.15(a). They were almost certainly launched by Ampère forces. Another feature of the photograph is a ball of ultraviolet light which was excited remotely in the air molecules by electromagnetic radiation from the arc.

That this ball of light was not initiated by a plasma sphere was easily proved by placing strong ultraviolet and color filters in front of the camera. The shape of the plasma jet, which emerged through the stack of filters, is shown in figure 6.15(b). It was a vertical jet fountain above the arc, as drawn in figure 6.14. This jet was produced with a 2.5 μF capacitor charged to 20 kV, thus storing 500 J. The maximum current of the underdamped discharge was 22 kA.

The phenomena underlying retrograde lightning strokes were investigated in NG's Oxford laboratory, the HCS laboratory in Toronto and the TCBOR laboratory at Richmond VA, the latter under the direction of Richard Hull. Water plasma accelerators, based on the working principle outlined in connection with figure 6.13, were employed to accelerate cold, but ionized, water mist or fog to high vertical velocities above the muzzle of the accelerator. With energies stored in capacitors of 100 J and less, it was possible to accelerate the water droplets to supersonic velocities, as proved by the photograph of figure 6.16. The conical tip of the jet of water mist is formed by ablation in still air and establishes the progress of a shock front travelling at a velocity greater than the velocity of sound in air (340 m/s). In the subsonic range of speeds, the tip of the jet of water droplets was blunt and mushroom shaped.

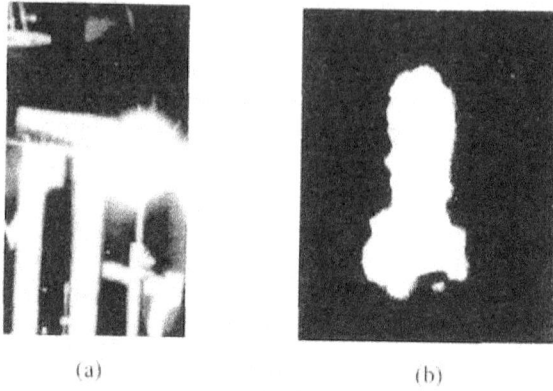

(a)   (b)

Figure 6. 15 (a) Ultraviolet ball of light surrounding the arc gap described in figure 6 14
(b) The same arc as (a) but viewed through a stack of colored filters

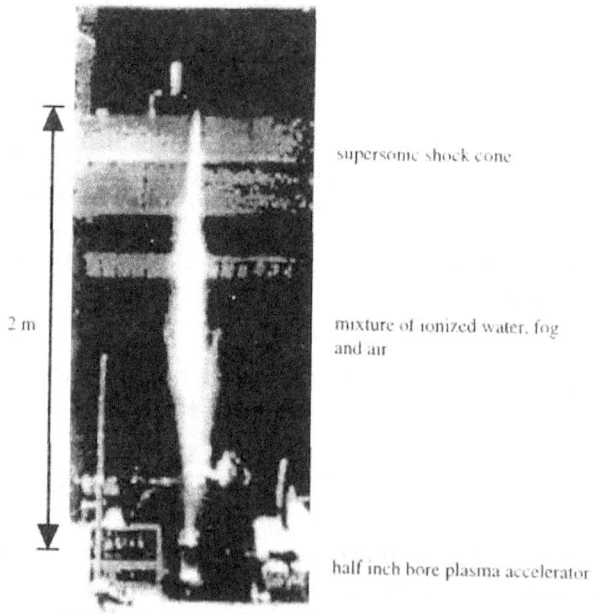

supersonic shock cone

2 m

mixture of ionized water, fog
and air

half inch bore plasma accelerator

Figure 6.16 : Two meter tall ionized fog column produced by exploding 1 ml of water
with a 225 J electric arc (courtesy of Richard Hull)

The performance of the laboratory plasma accelerator may be compared with retrograde lightning on the basis of the action integral of the current pulse, which controls the acceleration force and the initial momentum of the plasma jet. In some of the supersonic laboratory shots the action integral was less than 500 A^2s. The highest reported action integral for a lightning stroke was 22×10^6 A^2s [6.19]. Hence, very large initial ion velocities may be expected in retrograde lightning.

Above the thunder cloud the plasma will be a mixture of air and water mist droplets. Air alone has also been accelerated in laboratory devices. The momentum imparted to the air plasma alone is much smaller than that observed with air and fog mixtures. The energy involved in plasma explosions will be treated in greater detail in the last chapter.

Chapter 6 References

6.1   J.D. Cobine, *Gaseous conductors*, Dover, New York, 1958.

6.2   M.F. Hoyaux, *Arc physics*, Springer, New York, 1968.

6.3   H. Aspden, "Anomalous electrodynamic explosions in liquids", IEEE Transactions in Plasma Science, Vol.PS-14, p.282, 1986.

6.4   P. Graneau, N. Graneau, "Electrodynamic explosions in liquids", Applied Physics Letters, Vol.46, p.468, 1985.

6.5   I. Gilchrist, B. Crossland, "The forming of sheet metal using underwater electrical discharges", IEE Conference Publications No.38, p.92, Dec. 1967.

6.6   F. Frungel, (a) "Zum mechanischen Wirkungsgrad von Flussigkeitsfunken", Optik, Vol.3, p.125, 1948; (b) *High speed pulse technology*, Vol.1, p.477, Academic Press, New York, 1965.

6.7   R. Azevedo, P. Graneau, C. Millet, N. Graneau, "Powerful water-plasma explosions", Physics Letters A, Vol.117, p.101, 1986.

6.8   L. Dragone, "Electric arc explosions: A thermal paradox", Journal of Applied Physics, Vol.62, p.3477, 1987.

6.9   P. Graneau, P.T. Pappas, L.J. Ruscak, D.W. Swallom, "Electrodynamic water arc gun", 4th Symposium on Electromagnetic Launch Technology, University of Texas at Austin, April 1988.

6.10   W.J. Remillard, "The history of thunder research", Weather, Vol.16, p.245, 1961.

6.11   P.E. Viemeister, *The lightning book*, MIT Press, Cambridge MA, 1972.

6.12   P. Graneau, "The cause of thunder", Journal of Physics D: Applied Physics, Vol.22, p.1083, 1989.

6.13   M. Hirn, "The sound of thunder", Scientific American, Vol.59, p.201, 1888.

6.14   W. Schmidt, "Ueber den Donner", Meteorologische Zeitschrift, p.487, 1914.

6.15   V.I. Arabadzhi, "On the characteristics of thunder", Dokl. Akad. Nauk, Vol.82, p.377, 1952.

6.16 R.D. Hill, "Thunder", *Lightning* (R.H. Golde, Editor), Vol.1, Academic Press, New York, 1977.

6.17 M.A. Uman, *Lightning*, Dover, New York, 1984.

6.18 R.E. Orville, "Lightning spectroscopy", *Lightning* (R.H. Golde, Editor) Academic Press, New York, 1977.

6.19 K. Berger, "The earth flash", *Lightning* (R.H. Golde, Editor), Academic Press, New York, 1977.

6.20 W.J. Broad, "New class of lightning found high above the storm clouds", The New York Times, January 17, 1995.

6.21 K. Davidson, "Bolts from the Blue", New Scientist, Vol. , p.32, Aug.19, 1995

Electrodynamics in the Quest for New Energy

New Energy

The inevitable exhaustion of fossil fuels in the next few centuries provides major incentives for finding new sources of energy to maintain and advance our civilization. The ideal solution would seem to be renewable energy. Examples are hydroelectric and wind power, as well as the direct utilization of solar radiation. At first sight it appears that electrodynamics is not a factor in the development of renewable energy sources.

Forty years ago it was thought that large atomic power plants would satisfy the need for electricity generation while fossil fuels were still plentiful. This hope has faded because of the unfavourable environmental impact of fission reactors. Governments of the industrial nations then switched their support to controlled thermonuclear fusion research and in particular to Tokamak reactor development. This is expected to be environmentally benign, but research on these machines has run into a string of largely electrodynamic difficulties which dim the prospect of controlled thermonuclear fusion.

Individual inventors and privately funded scientists are now trying to find new energy in small scale experiments which do not depend on massive funding from governments and huge corporations. The best known effort in this direction is 'cold fusion' which was discovered in 1989 in the United States at The University of Utah. It has become a hotly debated science, driven by the controversy itself, and feared by the large teams working on Tokamak reactors.

The great number of papers published in the cold fusion area leaves little doubt that excess heat has been produced in a number of different experiments. The problem remains how to make the process reliable and economical.

The phrase 'cold fusion' could also be applied to several filament fusion techniques which have been researched for as long as thermonuclear reactions. Plasma filament fusion processes operate at much higher temperatures than recent cold fusion experiments, but far below the one hundred million degrees required in the toroidal chambers of Tokamaks. The Newtonian electrodynamics plays a prominent role in filament fusion experiments, and this will be discussed further in this chapter.

Only recently have we realized that the anomalously large forces arising in water arc

228

plasma explosions are probably derived from energy residing in ordinary water which may, one day, be harnessed for technological ends. Electrodynamic action seems to be essential for liberating the anomalous water energy. This potential energy could be of nuclear origin, or it may even be the free vacuum energy which, according to quantum field theories, should pervade all space, however we are inclined to speculate that it is intermolecular bonding energy depending on the molecular structure of water. The last topic to be analyzed in this book is the electrodynamic liberation of anomalous energy in water plasma explosions.

Fusion Research in 1995

Thermonuclear fusion has been researched since the early 1950s. Much of the effort was devoted to magnetically confined toroidal Tokamak plasma reactors. This type of machine requires complex and very costly external electromagnets. Conservative estimates suggest that two or three more generations of large advanced Tokamak reactors have to be designed, built and tested, over a period of twenty to thirty years, before commercial electricity generation by nuclear fusion becomes viable. It is generally agreed that enormous resources from all industrialized nations will have to be pooled to complete this task.

Against this background, it is surprising to find that research on far simpler fusion schemes, not depending on plasma confinement by magnets, are being discouraged by government agencies, and in many instances have been cut off from funding.

In this chapter we consider three non-thermonuclear fusion processes which are closely related to each other and could be collectively classified as 'filament fusion'. In all three methods large current pulses have been passed through short (~10 cm) and thin (~1 mm diam.) gaseous, liquid, or solid filaments. The neutron yield from deuterium filament fusion reactions has been as high as 10^{12} per shot.

Gaseous filament fusion has been achieved with 'plasma focus' devices. Liquid filaments were used in 'capillary fusion'. Dense z-pinch experiments have been performed with solid filaments. Today researchers face the challenge to design further small-scale filament fusion experiments in which the reaction rate, or neutron emission, is increased by several orders of magnitude to achieve energy break-even and ultimately a net gain in useful energy.

The search for new small-scale devices must inevitably be guided by the understanding of what makes filament fusion work. There is no agreement on this issue. Some investigators continue to believe that plasma pinch and MHD instabilities create highly local, very small plasma regions in which the temperature rises to levels at which thermonuclear fusion can occur. Others think conventional electrodynamic Lorentz forces are somehow involved in the acceleration of nuclei to sufficient velocity so that they may fuse on impact with each other. Based on the Newtonian electrodynamics described in this book, we will examine the prospect of aligned fusion collisions in which the nuclei are accelerated by longitudinal Ampère forces.

MHD Instabilities

Quite early in fusion research, many neutrons were produced with capacitor discharges through low-pressure deuterium gas. Deuterium is an isotope of hydrogen, and the nucleus of this isotope is called a deuteron. It contains a neutron in addition to the proton of ordinary hydrogen. There was no doubt that in the discharges through deuterium the neutrons arose from deuteron-deuteron (D-D) fusion reactions. The neutron yield was, however, too great, and the energy input into the plasma too small, to account for the neutrons by the heating of the plasma and thermonuclear reactions. This was in fact the first cold fusion experiment reported by Anderson et al [7.1]. Rather than being hailed as an important discovery, it was labelled a failure because the reaction could not set up a self-sustaining nuclear 'burn' which was thought to be an essential requirement of a commercial fusion reactor.

An analysis of neutron velocities and directions of flight revealed information about the deuteron motions prior to the fusion collisions. In this way it was found that the nuclei had not been in randomized thermal motion, but travelled at high velocity along the current streamlines. Furthermore, the axial deuteron velocity prior to collision appeared to have been so large that it could not have been produced by the acceleration due to the local electric field. This electric field was set up by the potential difference maintained between the electrodes at the ends of the plasma column. The plasma current generated an azimuthal magnetic field, which gave rise to a pinch action, but it could not explain the axial particle acceleration. In the 1950s the investigators of the fusion phenomenon knew nothing about longitudinal Ampère forces which have since explained their observations.

To understand the dynamics of plasmas it is necessary to combine the electrodynamic equations with those of fluid dynamics and thermodynamics. This leads to the magneto-hydrodynamic (MHD) equations. They involve the atomic or molecular species of the gas, the degree of ionization, particle density and initial pressure and temperature.

In this complex MHD-theory the plasma behavior is strongly influenced by the Lorentz force. This is given by

$$\vec{F}_L = q\,(\vec{E} + \vec{v} \times \vec{B}) \qquad (7.1)$$

in which q is the electric charge of the accelerated particle. $\vec{E}$ is the electric field strength and $\vec{B}$ the magnetic flux density at the location of the charge, and $\vec{v}$ is the relative velocity between the charge and the observer.

We will look at a cylindrical plasma column of length d between a pair of electrodes with an applied voltage V. This was investigated by Anderson et al [7.1] in the first linear deuterium pinch experiment. The electric field strength in Eq.7.1 is then given by

$$\vec{E} = \frac{V}{d} \qquad (7.2)$$

The important particle in these pinch experiments was the deuteron which has a positive charge of 1.6×10^{-19} C and a mass of 3.34×10^{-27} kg. The deuterium was contained in a glass tube, and the magnetic pinch contracted the plasma column diameter away from the glass enclosure.

It is an empirical fact that the cross-section which a deuteron presents to another colliding deuteron is not constant, but varies with the relative velocity of the deuterons. Until quite recently it was argued that the so-called collision cross-section of deuterons was negligibly small for deuteron energies of less than 10 keV. Since 1 eV = 1.6×10^{-19} J, the kinetic energy of a 10 keV deuteron involves a linear velocity of approximately 1000 km/s. This is the relative velocity with which the two deuterons must collide to have a finite chance of producing a fusion reaction.

In the Anderson et al experiments the electric field strength was about 100 kV/m, and the deuteron pressure of the order of 1 torr. The mean free path between deuterons was therefore less than 1 mm. Over this path deuterons could acquire, on the average, no more than 100 eV of energy or a relative velocity of 100 km/s. This was considered to be far too small to cause impact fusion reactions.

It leaves us to consider the magnetic component of the Lorentz force, Eq.7.1, as the cause of nuclear collisions. This force is everywhere perpendicular to the current streamlines and produces the pinch effect, reducing the plasma column diameter. In a column of constant length the pinch action is unable to generate the required axial velocities. Hence the Lorentz force fails to explain the large axial velocities of deuterons observed in the Anderson et al experiments.

The same conclusion emerged when the full MHD-equations were taken into account. But then it was noticed that the magnetic pressure on the plasma column was in unstable equilibrium with respect to the thermal expansion forces. A small disturbance in particle motion would upset the equilibrium of the cylindrical plasma column and distort it in one of several possible modes. The lowest order and most common mode of distortion was a necking down of the column diameter in one or more places. This was given the name of 'sausage instability', for it makes the column look like a string of sausages.

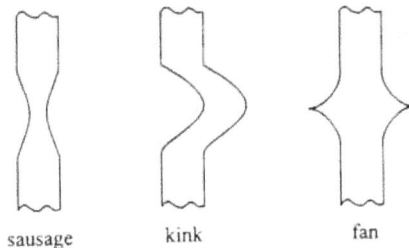

| sausage | kink | fan |

Figure 7.1 : Instabilities of a plasma column

Figure 7.1 illustrates three different types of plasma column instability. Both the sausage and the kink deformation concepts arose from MHD-theory. The fan instability should result from longitudinal Ampère forces. In the Ampère electrodynamics all current elements in a straight filament repel each other. Thus assuming that the electrodes are prevented from moving apart, the filaments will tend to buckle. Since they constrain each other from moving inward, they will all tend to spread laterally outward. This deformation could not take place in a solid metal conductor because of inter-atomic bonds. The bonds, however, are absent in fluids and plasmas.

Let us examine the sausage instability in more detail. The current is carried almost exclusively by electrons, and only these fast moving electrons are pinched inward by the Lorentz force, Eq.7.1. Therefore the outer layers become partially depleted of conduction electrons. The positive ions of the outer layers are then attracted to the axis of the plasma column by Coulomb forces. It is this inward attraction of ions which gives rise to the pinch contraction. The effect is particularly strong in the sausage neck. Since the pinch force turns out to be an inward attraction, rather than an external magnetic pressure, the ions will move collectively toward the axis without many collisions. In this way the ions can acquire large radial velocities. Lochte-Holtgreven [7.2] has calculated that the collisions of counter-streaming deuterons across the axis are powerful enough to establish a thermally disordered plasma of sufficient temperature to cause thermonuclear reactions. In this way the sausage instability would set up a thermonuclear hotspot.

If true, this should result in uniformly distributed isotropic neutron emission. However as Anderson et al [7.1] have shown, with their pinch experiments, the neutrons travelled preferentially in the axial direction. Hence the experimental evidence contradicts thermonuclear hotspot formation at sausage instability sites.

Another possibility is that the deuteron concentration in the sausage neck leads to axial expansion of the deuterium gas in both directions away from the neck. This kind of expansion would have to be driven by randomly directed collisions. If strong enough, this could lead to thermonuclear reactions. This explanation, however, is once more denied by the anisotropic emission of neutrons.

Most often the axial deuteron acceleration is wrongly attributed to induced electromotive forces. Starting with Neumann's generalized law of induction, Eq.4.11, we can show that for a finite length conductor of selfinductance L, and carrying a total current I, the induced e.m.f. over the length of the conductor may be written

$$ e = -\frac{d}{dt}(L\,I) = -\left(L\frac{\partial I}{\partial t} + I\frac{\partial L}{\partial t} \right) \tag{7.3} $$

When L is constant and I increases with time, the induced e.m.f. is negative. This means it is a back-e.m.f. opposing current flow and trying to keep the current constant. When I is constant and L increases with time, the induced e.m.f. is again negative and opposing current flow. Now when the conductor diameter contracts due to radial pinch forces, its selfinductance increases. Hence, over the length of the pinch neck, Neumann's law predicts the acceleration of positive charges -- that is deuterons -- in the direction opposite to current flow.

The negative sign of Eq.7.3 has often been overlooked, as for example in the paper by Anderson et al [7.1]. If the deuteron flow, caused by the neck formation, cancelled part or all of the current flow, the pinch force on the neck should relax and allow radial expansion. Hence the dL/dt effect at sausage instability sites is not necessarily as large as has been speculated.

In the MHD instabilities the pinch forces overcome outward directed thermal pressure of the plasma column. Both Ampère's and Lorentz's electrodynamics predict the same pinch forces. Hence when an Ampère MHD-theory is developed, it will probably predict sausage and kink instabilities in addition to the fan instability. The problem with the Lorentz MHD has been that it could not explain any axial motion of the deuterons other than by thermal agitation. In respect to axial forces the Newtonian electrodynamics is decidedly more helpful.

Capillary Fusion

The announcement made by the University of Utah, in March 1989, that two professors had produced excess heat in a special electrolytic cell gave rise to the term *cold fusion*. Since then the general public has been given to understand that the only viable controlled fusion process is the thermonuclear one which proceeds at about one hundred million degrees. Even cold fusion scientists were not aware that fusion had been demonstrated far below the $10^{8\circ}$K level. In an otherwise excellent review of cold fusion in the MIT Technology Review [7.3], Storms made the statement:

> "Fusion had been known to occur only in the stars and thermonuclear bombs; attempts to harness it for energy had been limited to systems that heat hydrogen fuel to extremely high temperature using complex and expensive equipment."

This is not true. Non-thermal fusion research has been in progress for over forty years with support from the U.S. and other governments. The arguments which have been made for and against cold fusion almost all ignore the large body of published information on plasma focus fusion, solid deuterium fiber fusion and capillary fusion. The experiments by Anderson et al [7.1], made in the 1950s in the Berkeley Radiation Laboratory, laid the foundation of non-thermal filament fusion.

These experiments utilised magnetic pinch tubes of about 5 cm diameter and 40-100 cm length, which were filled with deuterium gas at the low pressure of one or two torr. It turned out to be surprisingly easy to produce a considerable number of neutrons by discharging a 50 kV, 12.5 µF capacitor bank through the pinch tube. The generation of 4×10^8 neutrons per pulse discharge was taken to be proof of having successfully collided and fused at least twice that number of deuterium nuclei. The enthusiasm which this early success aroused was dispelled when it was realized that the nuclear reactions were not caused by thermal collisions!

At the time it was reasoned that controlled fusion would never be commercially viable unless the fusion reactions ignited the deuterium and the plasma could be left to burn without supplying external energy to it. Anderson et al listed ten reasons why their nuclear reactions could not possibly have been of thermal origin.

The U.S. Program in Controlled Fusion was classified under the code name 'Project Sherwood'. Later the stricture of secrecy was lifted and Bishop [7.4], the former chief of the program, wrote:

> "Two bits of evidence were accumulated that could not be reconciled
> with the theory of thermonuclear origin. In the first place, the number of
> neutrons observed was too great; under the operating conditions of the
> experiments, the temperatures predicted from the Rosenbluth theory were too
> low to produce so many neutrons from fusion reactions. The second and even
> more convincing evidence was the result of a careful study of the energy
> spectrum of the neutrons which were emitted. This study carried out initially
> at UCRL, Berkeley, showed that while the neutrons were coming from the
> body of the discharge, the deuterons responsible for their production (through
> D-D reactions) were unquestionably moving with rather high velocities in the
> axial direction. The deuterons, therefore, did not have random velocities, as
> required for true thermonuclear conditions."

The Anderson et al [7.1] results for a particular discharge shot are reproduced in
figure 7.2. During the first quarter-cycle of the oscillating discharge current the amplitude
approached 400 kA. Neutrons were emitted in two short bursts before the peak current was
reached. The bursts lasted between 100-200 ns, while one cycle of the current took 8 μs.
Hence the fusion collisions were obviously caused by some plasma instability phenomenon
which took time to develop, and then destroyed the condition for further neutron emission,
although the current continued to flow through the plasma. Ten years later this strange
behavior led to the realization that there might be a connection between wire explosions and
filamentary fusion.

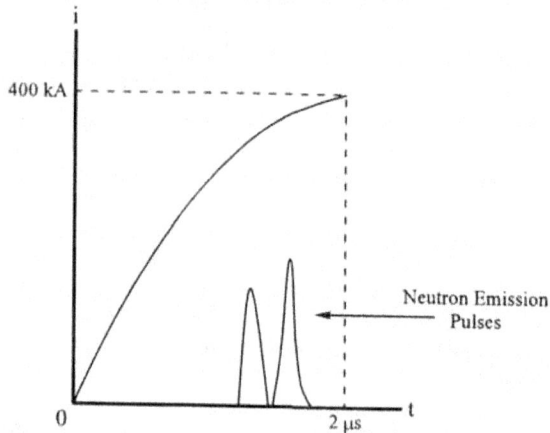

Figure 7.2 . Current and neutron pulse emission during the first quarter cycle of
the ringing discharge current [7.1]

Lochte-Holtgreven [7.2] and his collaborators at the University of Kiel were familiar with the nature of wire explosions and applied this knowledge to a process which they called *capillary fusion*. They may also have known that the gaps between wire fragments are preserved in the later liquid and vapor phases of the explosion. Nasilowski [2.4] had performed experiments in which a wire was buried in sand and the current pulse through it lasted long enough to vaporize the metal. He then X-rayed the sand and found the metal vapor pattern shown in figure 7.3. He called this wire striations. Apparently the wire was fragmented into solid pieces which were roughly the length of the wire diameter. Later these pieces melted and evaporated with the current still flowing through air-arc gaps. The vapor expanded to three times the wire diameter without closing the gaps between the striations. Once the current pulse was over, the metal vapor condensed on nearby grains of sand, freezing the striation pattern of figure 7.3.

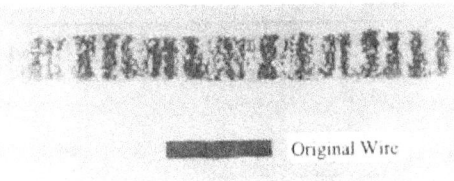

Original Wire

Figure 7.3 : Nasilowski's copper vapor striations in sand [2.4]

The Kiel group published their most important paper on capillary fusion in 1976 [7.2]. It reported experiments with a solution of lithium in heavy ammonia $Li(ND_3)_4$ consisting of 70 atomic percent of deuterium. The capillary filaments were 7 to 8 cm long, and from 0.5 to 1.5 mm in diameter, set in a block of glass which was compressed with dry nitrogen at 10 to 20 atm.

Current was forced through the conducting liquid filament by a capacitor discharge from a 5μF bank charged to voltages between 100 and 200 kV (100 kJ maximum). The discharge circuit was underdamped which resulted in current oscillations of about 200 kHz. Bursts of 10^4 to 10^5 neutrons were detected in every shot with voltages between 150 and 200 kV. When light ammonia (no deuterium) was substituted for heavy ammonia, the discharge produced no neutrons. This was taken to be positive proof of fusion reactions in the heavy ammonia solution.

The neutron bursts lasted for 30 to 50 ns. Just as in figure 7.2, they were much shorter than the ringing capacitor discharges of 10 μs duration. Lochte-Holtgreven did not recognize the parallel between his experiments and the earlier ones carried out with pinch tubes in Berkeley. This explains why he continued to search for a thermonuclear explanation of capillary fusion.

In the Kiel experiments, each neutron burst occurred at the same time as a dip in the current oscillogram. This dip was almost certainly caused by the disruption of the liquid plasma filament. At the time of the neutron burst, the current had risen to only about ten percent of its maximum value of 10 kA. This is indicated in figure 7.4.

Another important observation made at Kiel University was the fracture of the glass block which followed each neutron burst with a delay of 100 to 300 ns. It seems unlikely that the glass was broken by thermal forces, because information in the paper [7.2] can be used to show that most of the electrical energy was converted directly to mechanical energy, via an induced back-e.m.f., without passing through heat. The significance of this observation was not recognized in 1976, and it is important that it should be confirmed with additional experiments.

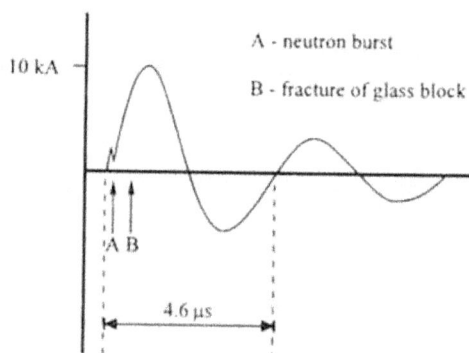

Figure 7.4 : Discharge current oscillogram during a capillary fusion event [7.2]

The energy conversion finding comes from the oscillogram of figure 7.4. In this case the 5 μF capacitor bank was discharged from 100 kV. Dividing the applied voltage by the first current peak indicates a plasma filament impedance of the order of $Z = 10 \, \Omega$. This is unusually large for an experiment of this nature. Ignoring the ionization process, it would normally be assumed that the total impedance, Z, is the sum of the surge impedance, Z_0, and the resistance, R, of the circuit, represented by

$$Z = Z_0 + R \qquad (7.4)$$

With L and C being the circuit selfinductance and capacitance respectively, we can write

$$Z_0 = \sqrt{\frac{L}{C}} \qquad (7.5)$$

The oscillating frequency, f, of the series circuit is given by

$$f = \frac{1}{2\pi\sqrt{LC}} \qquad (7.6)$$

This frequency was measured to be $f = 217$ kHz. Using Eq 7.6 indicates that $L = 0.1$ μH. Lochte-Holtgreven correctly described this as a low-inductance circuit.

The lithium-ammonia solution was said to have had a resistivity similar to that of liquid mercury. For a filament of 8 cm length and 0.5 mm diameter, this results in a resistance of $R = 0.4$ Ω. After the liquid has been ionized it will be a better conductor. Thus 0.4 Ω is an upper bound of the filament resistance.

Using Eq.7.5 the surge impedance, Z_0, comes to 0.14 Ω, and thus with Eq 7.4, the total circuit impedance comes to $Z = 0.54$ Ω, which was much less than the observed 10 Ω. This discrepancy leads to the following conclusion.

Z_0 in Eq.7.4 allows for the storage of magnetic energy during the first current rise of figure 7.4. R accounts for the generation of Joule heat. Equation 7.4, however, ignores the kinetic energy gained by the deuterons as well as the energy needed to break the glass block before the first current peak in figure 7.4 is reached. The last two observations can be regarded as representing mechanical work. For this purpose the back-e.m.f. e_b has to be induced in the circuit. This opposes the instantaneous current i. If e is the instantaneous driving e.m.f. applied between the ends of the capillary filament, then the instantaneous electric power supplied becomes

$$e\,i = i^2 Z + e_b\,i \qquad (7.7)$$

or

$$\frac{e}{i} = Z + \frac{e_b}{i} \qquad (7.8)$$

Since $e/i = 10$ Ω (observed) and $Z = 0.54$ Ω (deduced), we have $e_b/i = 9.46$ Ω, indicating that nearly 95 percent of the electric energy consumed was converted to mechanical energy. Therefore capillary fusion experiments appear to be an extraordinarily efficient method of accelerating nuclei, unless much of the energy is wasted in breaking the glass tube.

Haendel and Jonsson [7.5] reviewed the Kiel experiments. The first explanation of capillary fusion in terms of Ampère forces was published in 1992 [7.6]. If the high measured impedance can be confirmed by other investigators, and longitudinal Ampère forces were responsible for the acceleration of nuclei and glass breakage, one way to improve the energy utilization would be to use much stronger capillary tubes, leaving more of the mechanical energy for particle acceleration. On the other hand, if capillary fusion is mainly due to pinch forces, stronger capillaries will make little difference to the neutron yield. Hence the issue of the nature of the deuteron acceleration forces can be resolved by experiment.

In Chapter 2, it was shown that Ampère's force law predicts that the minimum

238   *Electrodynamics in the Quest for New Energy*

fragment length of an exploding square cross-section wire is 1.4 times the width of the conductor [7.7]. From this we may expect a cylindrical wire to break into fragments of a length of one to two conductor diameters which generally agrees with experience. Disrupted plasma columns would be expected to display the same minimum bead length to diameter ratio.

There is evidence in the literature confirming the break-up of a plasma filament into beads. Sethian et al [7.8] published an X-ray pinhole photograph of the disintegration of a deuterium fiber plasma filament. This was obtained during the evaporation of a 125 μm diameter solid deuterium fiber by a 350 kA current pulse of 130 ns rise time, and clearly shows the break-up into beads, none of which are shorter than the diameter of the fiber. In other experiments, Sethian et al [7.9] have proved that the X-ray emission coincides in time with the neutron emission and the onset of rapid radial expansion of the plasma filament, just when the current pulse reaches its maximum amplitude. For this investigation they reported the formation of 8 to 10 beads randomly spaced along a 4 cm long plasma filament.

The Newtonian electrodynamics not only explains bead formation but also the separation of the beads subsequent to formation [7.7]. For this purpose consider two blocks of 5×5×6 cubic current elements with an axial separation of g elements, as shown in figure 7.5. Here it must be remembered that the Amperian current element is not the travelling electron, but the magnetically polarized ion. Consequently, in the Newtonian electrodynamics, the electrons between the beads do not contribute to the electrodynamic forces. The specific repulsion force f_R between the upper and lower block was calculated when the current I = 25i flowed vertically. In this case f_R was the specific vertical force summed over all of the elements of either the upper or lower cell. Finite element computations yielded the results plotted on figure 7.6. The graph reveals that f_R is relatively constant for 0<g<5. The force actually increases a little with g when the gap is very short. This aids the disruption of the filament. Hence longitudinal Ampère forces readily account for beading and subsequent bead separation.

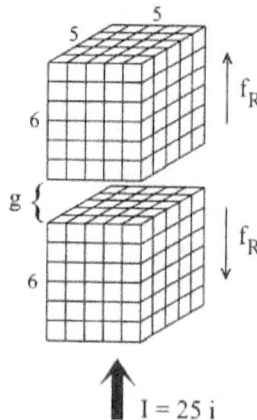

Figure 7.5 : Repulsion between adjacent cells of the plasma column [7.7]

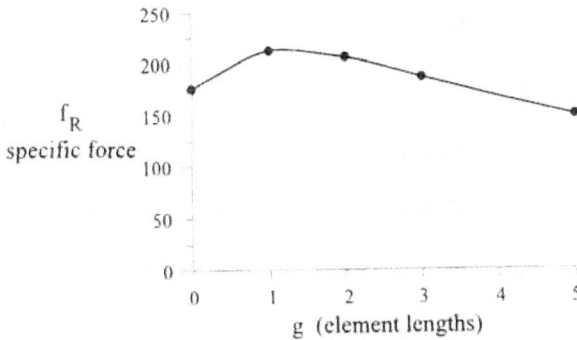

Figure 7.6 : Specific repulsion force between beads as a function of gap length

If filament fusion is the result of head-on collisions of deuterons, it must be considered how Ampère forces could achieve this goal. There exists the possibility that positive ions are repelled from opposite faces of a rupture gap and then collide in mid-gap. However previous calculations on short cells have revealed that this is unlikely when the column has been disrupted into many short pieces because the ions in the fracture faces are attracted back to their own cell. However in the early stages of filament break-up, when the fragments are still relatively long, mid-gap ion collisions could occur and they are likely to be sufficiently energetic to cause fusion. However if the gap ion density is too low to allow many such collisions, the ions will be decelerated by Ampère forces due to repulsion by the fragment on the other side of the gap.

Thus a more probable cause of fusion collisions is the compression of a small bead, due to repulsion forces from adjacent beads on both sides. Adjacent metallic electrodes would produce the same effect. Remembering again that all the Amperian current elements reside in the plasma beads, we examine the magnitude of the Ampère forces on the current elements (deuterons) of a bead due to interactions with the two adjacent beads as well as all of the elements in its own bead, except itself.

Figure 7.7 refers to this problem. It involves three 5×5×6 adjacent cells with g = (1 element length) gaps between them and examines the forces on the elements in the central one. Other cells in either direction of the plasma filament, beyond those shown in the diagram, will further increase the compressive forces in the target cell, but their influence decreases with the inverse square of the distance.

To simplify the numerical treatment, only the forces on a single filament of the middle cell are calculated, as indicated in figure 7.7. In figure 7.7(a) this is the central filament, and in figure 7.7(b) it is a corner filament. The results which would be obtained for other filaments are intermediate between these two extremes. The finite element results for these two filaments are listed in table 7.1.

In table 7.1 a positive sign represents an upward directed force, thus indicating that both filaments are subject to axial compression, with the corner filament experiencing less stress. The forces in the central filament on elements 1 and 6 predict that the deuterons in these elements will be accelerated inward. The opposing motions from the two ends of the cell should produce collisions in the central elements 3 and 4. This argument ignores collisions due to thermal motion of the nuclei.

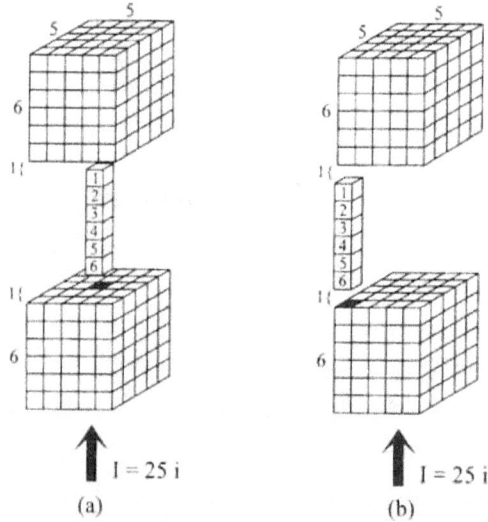

Figure 7.7 : Finite element analysis of compressive forces in the middle plasma bead of three adjacent beads [7.7]

m	Center filament	Corner filament
1	-1.39	0.035
2	-0.101	-0.069
3	0.096	-0.010
4	-0.096	0.010
5	0.101	0.069
6	1.39	-0.035

Table 7.1 : Specific force calculations for figure 7.7. (Positive force is upward)

The specific forces of table 7.1 can be interpreted in terms of deuteron accelerations by considering a numerical example. If the plasma column of 5×5 element cross-section carries a total current of 250 kA, then the current in each filament is 10,000 A. For a specific force of 0.75, which is intermediate between the forces on the first two elements of the center filament, the inward directed element force becomes

$$\Delta F = 0.75 \left(\frac{\mu_0}{4 \pi} \right) i^2 = 7.5 \text{ N} \qquad (7.9)$$

In filament fusion experiments the deuteron density is typically 10^{25} m^{-3}. If the plasma has a cross-section of 1×1 mm^2, each element will have a volume of 8×10^{-12} m^3. Hence the number of deuterons in one element is 8×10^{13}. If this number shares the force equally, then each deuteron is subject to a force of 9.4×10^{-14} N.

If the distance over which this force remains approximately constant is the length of one finite element (0.2 mm), then the energy acquired by the deuteron becomes 1.9×10^{-17} J, which equals 117 eV. In head-on collisions twice this energy is involved and we may say that 234 eV is available for producing fusion reactions. Of course, if the force is not shared equally by all deuterons in the volume element, then the collision energy could be much greater. Apart from the position of a deuteron in the plasma column cross-section, the energy it can acquire from Ampère forces will also depend on the length of the adjacent beads and the distance from the solid electrodes.

The question of how the longitudinal Ampère forces divide themselves between the deuterons of the fusion filament is clearly unresolved. It requires further investigation, preferably by experiment.

Finally, it must be noted that the reduction of the Coulomb barrier potential between colliding nuclei by electron screening and tunnelling phenomena, as discussed by Rambaut [7.10], is of the utmost importance. The calculations presented in this paper suggest that certain electromagnetic factors permit fusion reactions to occur at lower particle energies than assumed in thermonuclear fusion. Rambaut has argued that these phenomena will be particularly important in the case of capillary fusion.

The observed emission of neutrons at the instant of bead formation leaves little room for doubt that Ampère forces are the crucial aspect which makes capillary fusion possible.

Plasma Focus Fusion

A filament fusion process, which received far more support than capillary fusion, is known as plasma focus fusion. Governments in the United States, Britain, Italy and Germany have funded research in this area for almost as long as controlled thermonuclear fusion has been researched. The electrodes of a plasma focus device are shown in figure 7.8. This assembly has to be enclosed in a vacuum vessel in which the low deuterium pressure of 3-4 torr can be maintained. Plasma focus experiments are a development of the pinch tube work by Anderson et al [7.1] which dealt with non-thermal nuclear fusion. A new feature of the plasma focus geometry was the very dense deuterium filament created by magnetic pinch forces. This was responsible for describing the process with the term "focus".

When discharging the high voltage capacitor bank through the dilute deuterium gas, the first flashover occurs across the surface of the insulator which separates the central anode from the cathode tube. Electrodynamic forces sweep the resulting arc upward. As the discharge passes the tip of the anode, the plasma filament starts to form, as shown in figure 7.8. Depending on the voltage and capacitance of the current source, the filament may be one or more centimeters long. Its diameter varies during the course of the discharge, but is usually not less than one millimeter.

A large number of neutrons are released from the filament as a result of fusion reactions. Neutron emission is limited to a brief period of time during the current pulse, just when the filament is known to rupture. The neutron yield has in some experiments approached 10^{12} per shot. It has never been claimed that the entire filament reaches thermonuclear reaction temperatures. Much attention has been directed on local effects along the plasma filament and particularly the regions where the ruptures occur.

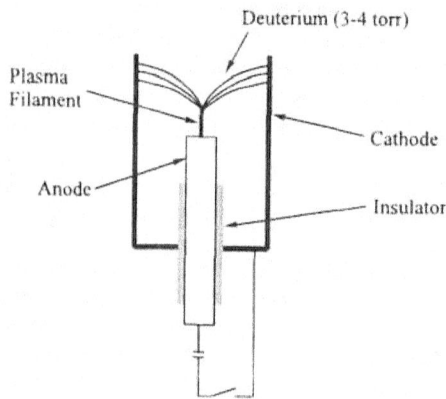

Figure 7.8 : A typical coaxial plasma focus device

As in earlier pinch tube experiments, the neutron output does not depend on the voltage applied between the ends of the filament. It is governed, primarily, by the pulse current amplitude. These two observations suggest that the deuterium acceleration forces are of electrodynamic rather than electrostatic origin. In a 1981 review, Haines [7.11] points out that in four different plasma focus experiments, carried out in two separate laboratories, the neutron yield was proportional to the fourth power of the maximum current, which ranged from 0.3 to 1.1 MA.

With optical framing photographs of 5 ns exposure and time resolved neutron detection, Decker and Wienecke [7.12] have proved that the neutron emission always coincides with filament rupture. The appearance of such an event is shown in figure 7.9. There is no doubt that pinch forces are responsible for the formation of the plasma filament, but we are faced with two possible rupture mechanisms. Without knowledge of longitudinal Ampère forces, investigators had no choice and attributed filament rupture to an m = 0 MHD sausage instability. This instability forms a neck in the filament, and the consequent radial current components on both sides of the neck and the resulting longitudinal repulsion between the radial currents may fracture the plasma column. Conversely, using the Newtonian electrodynamics would predict the longitudinal repulsion of current elements (deuterons), given by Ampère's force law, Eq.1.24, to be the cause of filament rupture without neck formation (see figure 7.9). In both cases the electron current continues to flow across the gap without producing a visible plasma.

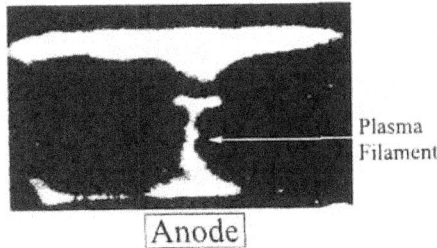

Figure 7.9 : Plasma focus filament disruption [7.11]

The opening of the rupture gap (figure 7.9) requires axial ion motion. Haines [7.11] mentions that there is also a center-of-mass motion away from the central electrode (anode). The simplest explanation of this motion is again longitudinal Ampère repulsion between current elements in the anode and others in the plasma filament. Very high ion velocities, of the order of 100 km/s, have been quoted in the literature.

The neutrons are certainly not produced by thermonuclear reactions because their flux is anisotropic, with most of the neutrons flowing along the filament axis. Indeed the fusion reactions appear to be a consequence of axially accelerated deuterons. The mechanism which has been put forward most frequently to account for the axial ion motion relies on ion acceleration by high voltage across the $m = 0$ neck [7.1]. It has already been shown in conjunction with Eq.7.3 that this is wrong because of a sign error in the induced e.m.f. equation. This leaves filament rupture by longitudinal Ampère forces as the only viable explanation, and must be the same mechanism which leads to capillary fusion.

Very large current pulses, in the mega-ampere range, decrease the effectiveness of plasma focus devices. Investigators at the University of Stuttgart and Imperial College, London, are responsible for the following pessimistic outlook [7.13], based on their experience with the most powerful plasma focus devices. They said:

"Particularly in the large PF (plasma focus) devices, however, it was found that the neutron yield stagnates or even decreases when the energy input and the current are increased above a certain critical value, despite extensive efforts to optimize the electrode dimensions. This effect seemed to limit the future of the PF as a fusion device."

The decline in neutron yield with current strength appears to be an indication of longitudinal Ampère forces becoming too powerful and that they disperse the plasma by radial expansion. This expansion would be the result of column instability which converts axial compression to radial expansion as in the "fan" instability described in figure 7.1. To overcome the fusion stagnation effect it may be advantageous to enclose the filament in a strong containment tube. This approach is supported by the following arguments.

The particle density in plasma focus filaments has been quoted as being 10^{25} m^{-3}. It is equivalent to the density of high-pressure electric arcs. Hence the knowledge gained in arc physics should apply to plasma focus experiments. In this respect the arc jets described by Sheer [7.14] are of interest. They have been produced with coaxial electrodes and various gas fillings. The coaxial electrode arrangement also works with arcs in liquids. An example is the water-arc gun shown in figure 6.8. It resembles the plasma focus device of figure 7.8, except that the insulator reaches up to the end of the central electrode, and the tubular electrode is a strong steel barrel.

Solid Fiber Fusion

Of all of the filament fusion concepts, the solid fiber experiments are most related to wire explosions. Tensile wire breaking is still the strongest evidence in support of the existence of longitudinal Ampère forces. Unfortunately, the deuterated polyethylene and solid deuterium filaments, which have been used in fiber fusion research, are insulators and not conducting wires. The initial current surge probably flows through a surface plasma raised from adsorbed gas on the dielectric fiber.

Lindemuth [7.15] has suggested that the fibers may not completely evaporate during the current pulses. The temperature of high-current arcs has been consistently overestimated, as already discussed in Chapter 6 in connection with lightning strokes. It is claimed that high temperature accounts for the radial expansion of the fiber plasma against the concurrent pinch forces during the fast current surge, as indicated by streak photography [7.8]. It could equally be the result of radial outward pressure generated by longitudinal Ampère forces.

The most important discovery made by Sethian et al [7.8] was that the neutron and X-ray emissions from the fiber plasma occurred at the moment of the longitudinal rupturing of the plasma column into a number (8 - 10) of beads, as revealed by the X-ray pinhole photograph of figure 7.10. This refers to a current of 350 kA, and the neutron yield was of the order of 10^7 per shot. For an 80 μm diameter deuterium fiber and currents ranging from 350 to 640 kA, the neutron production from D-D reactions scaled with the tenth power of current up to approximately 5×10^9 per shot at 640 kA.

From neutron time-of-flight measurements, Sethian et al deduced that the average deuteron was moving toward the cathode with an energy of 18 keV, or a velocity of 1300 km/s. Ion motion towards the cathode represents a positive current. It cannot be generated by the pinch induced e.m.f. (-i dL/dt) of Eq.7.3. Thus the only explanation of this high ion velocity appears to be longitudinal Ampère forces. These ions are probably responsible for at least some of the observed fusion reactions.

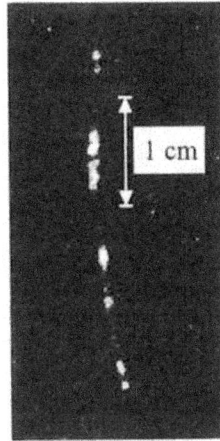

Figure 7.10 : X-ray pinhole photgraph of deuterium fiber fragmentation [7.8]

From their various measurements, and particularly from the stability of the plasma column, the researchers from the Naval Research Laboratory (Washington) [7.8] concluded:

"Clearly the neutrons do not come from a uniformly heated plasma. ....
These observations are obviously inconsistent with the predictions of MHD
theory and we need to look for features of the experiment that are not
included in the assumptions upon which the theory is based."

In subsequent experiments Sethian et al [7.9] raised the peak current to 920 kA, extended the pulse rise time from 130 to 840 ns, and increased the deuterium fiber diameter from 80 to 125 µm. These changes generally lowered the neutron yield which, at the highest current, was still only 4×10^9 per shot. Furthermore, the neutron count was no longer proportional to the tenth, but to the fifth power of maximum current. During the longer current pulses, neutrons were emitted in several bursts occurring over a period of several hundred nanoseconds. The neutron generation again coincided with the rupture of the plasma column in several places, as indicated by time correlated optical and X-ray photography.

As in the case of more conventional fusion devices, experiments with mega-amperes did not achieve the expected results. In all cases it has been a rapid radial expansion which foiled the experiments. This plasma expansion could be contained by enclosing the fiber in a strong capillary tube. If this were done the fiber might as well be replaced by a deuterium containing liquid. Keeping the fusible material away from containment walls is not important because of the relatively low temperatures occurring in this type of experiment. Hence fiber enclosure would bring us back to capillary fusion.

Does Cold Fusion Involve Capillary Fusion?

The application of static pressure to a confined volume of fusible matter has not produced nuclear fusion. In other words, it has been impossible to crush atoms with atoms. To bring two nuclei together, it appears that the atoms have to be stripped of all electrons. In the case of hydrogen and its isotopes this is quite easily done by the normal methods of ionization. On account of the attraction between electrons and positive nuclei, however, the electrons cannot be entirely removed from a collection of nuclei. The medium in which nuclear fusion will take place is, therefore, a mixture of nuclei and free electrons, that is a plasma.

Fusion then depends on overcoming the strong Coulomb repulsion between two nuclei by imparting to them high velocities which will make the nuclei collide. The conventional way of accelerating fusible nuclei has been to heat them until they produce thermonuclear reactions. In order to achieve this, temperatures of the order of 100 million degrees Kelvin are required.

Since the mid-1950s it has been known that fusion reactions can also be produced at substantially lower temperatures by accelerating them with what must have been electrodynamic forces [7.1]. The temperature in the pinch tube experiments was certainly less than $100,000°$ K and might have been as low as a few thousand degrees. This was the beginning of 'cold fusion'. In these early linear deuterium pinch experiments the neutron yield was far greater than expected. It was shown to be the result of organized (non-thermal) deuteron motion along the streamlines of current flow. The potential difference between the electrodes was quite small and could not have imparted sufficient velocity to the nuclei. Ten years later capillary fusion was discovered in which the accelerating potential was even smaller. These facts have led to the explanation of non-thermonuclear fusion processes by longitudinal Ampère forces.

The 1989 discovery and announcement by Pons and Fleischmann, at the University of Utah, of electrolytic cold fusion seemed to rule out fusion by energetic collisions of deuterons. The extremely low voltages combined with the short mean-free-path in the heavy water electrolyte and the palladium cathode, where the heat was evolved, should have made it impossible to accelerate the nuclei to significant velocities. In any case, the principal output of these cold fusion reactions was excess heat. The small number of neutrons and other fusion ashes raised the prospect of two energy generating mechanisms being present, with only one of them based on nuclear fusion. Filament fusion was not thought to be related because the electrolyte and the electrode metals did not look like they contained current filaments.

Then in 1991 the Los Alamos investigators Storms and Talcott-Storms [7.16] reported finding microscopic tubular channels in the palladium cathodes of cold fusion cells. Figure 7.11 shows their photomicrograph of a section through the capillary tubes in palladium. The tubes appear as round or elliptical dots. The photograph also displays one longitudinal section through a capillary cavity. The Los Alamos authors pointed out that a very high concentration of deuterium in the palladium cathode was required, and this was facilitated by the presence of the filamentary channels. The number of deuterium atoms absorbed in the palladium was found to be almost as large as the number of palladium atoms in the lattice.

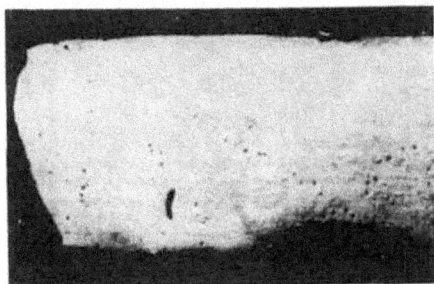

Figure 7.11 : Photomicrograph of palladium cathode material [7.16]

Two years later the Russian group of Kaliev et al [7.17] went further and actually provided capillary channels in a tungsten-bronze crystal of the structure shown in figure 7.12. This crystal had parallel channels perpendicular to the 100-face. The channels normally contained sodium atoms, however for the fusion experiments the sodium was replaced with deuterium atoms loaded into the crystal from a glow discharge of low pressure deuterium gas. When the 100-face was in contact with the anode or the cathode, the channels were all aligned with the electric field and capillary fusion reactions could take place in them. Apparently the crystal structure was strong enough to withstand any capillary explosions.

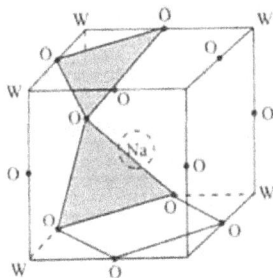

Figure 7.12 : Tungsten bronze crystal structure [7.17]

Some other cold fusion experiments were performed by Karabut et al [7.18] in Russia, however these researchers did not examine their palladium cathodes for capillary channels before proceeding with the experiments. They nevertheless found that the metal "wore out" and then contained a large number of microscopic bubbles in the "spent" state, as shown in figure 7.13. The bubbles were likely to have been the result of capillary explosions. In this investigation the deuterium was loaded into the palladium by a glow discharge.

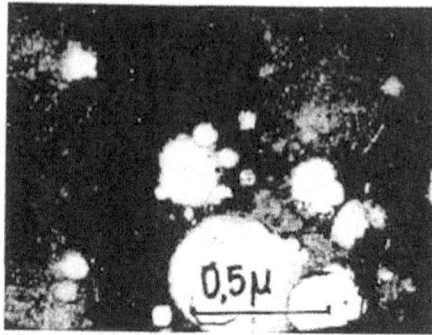

Figure 7.13 Transmission electron photograph of microscopic cavities in "spent" palladium cathode [7.18]

At the time of writing it is not understood why current in a metal should flow preferentially through capillary filaments of deuterium, unless the deuterium somehow remains ionized during its passage through the metal. The process would be reminiscent of the formation of superconducting filaments in a normal metal matrix. This aspect clearly requires further research.

Many experiments have revealed burst-like fusion reactions. They suggest that explosive events are involved, a fact which has been directly confirmed by Golubnichii et al [7.19]. These Russian investigators detected acoustic pulses which coincided in time and location with bursts of neutron emission from palladium cathodes. The most likely cause of the correlated emissions are capillary explosions.

The question arises of whether Ampère forces scale in such a way that capillary explosions can occur in the tiny channels found in palladium cathodes? In fundamental electromagnetic units (e.m.u.), the Ampère force (in dynes) is given by the product of the square of the filament current (in absolute amperes) and a dimensionless geometrical factor which is independent of the unit of length [1.12]. Hence for constant current, the force will remain the same when the size of the filament is reduced by a linear factor of 10^{-6}, from 1 mm diameter and 80 mm length, as in Lochte-Holtgreven's experiments [7.2], to 10 Å diameter and 800 Å length. Retaining the same number of finite current elements in both cases, the element volume will be reduced to 10^{-18} of its original value. Therefore, for the same density of deuterium atoms in both cases, and for the same force per deuteron, which is proportional to current squared, the current may be reduced by a factor of 10^{-9}. As the Lochte-Holtgreven experiment involved a current of 1000 A, in the palladium cathode only 1 μA per capillary filament may be required. This calculation indicates that cold fusion events could indeed be capillary explosions.

The two to five-fold energy multiplication achieved by Karabut et al [7.18] with palladium cathodes is very promising, if it can be independently confirmed. It indicated that metal can withstand capillary explosions better than a dielectric material like glass. Nevertheless, the capillaries in palladium were destroyed and the cathode material had to be replaced after only minutes of use.

It would seem worthwhile to explore fabricated capillaries in reinforced ceramic structures to see if energy gains from 200 to 500 percent can be obtained without destructive effects. Another reason for pursuing the dielectric route, in parallel with the metallic route, is the higher temperature at which ceramic components could be operated. For equal efficiency, the higher operating temperature would be an advantage in a fusion reactor in which heat has to be converted to electricity. Figure 7.14 shows a possible design of a reinforced capillary fusion cartridge.

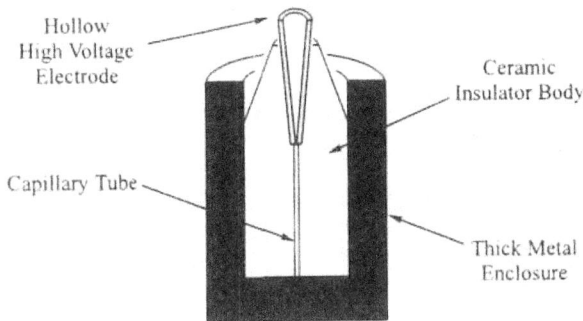

Figure 7.14 . Cross-section through proposed capillary fusion cartridge

Liberating Potential Energy from Water

1.   Anomalous Strength of Water-Plasma Explosions

Water arc explosions have been discussed in Chapter 6 as if they were purely electrodynamic events. The treatment outlined in that chapter was based on the experimental knowledge available at the beginning of 1994. Important new facts have come to light since then. The original assumption that the arc explosion was entirely propelled by electrodynamic forces no longer seems tenable.

On a number of previous occasions it has been explained that the signature of electrodynamic forces is their proportionality to the square of the current strength. To put it another way, the dimensionless k-factor of Eq.6.22 must be independent of the amplitude and shape of the current pulse. The four shots with the saltwater cartridge plotted in figure 6.6 are still approximately consistent with the general assumption of k being constant and independent of the action integral A. This guided our research on water plasma explosions for nearly a decade. Considerable efforts were made to discover additional electrodynamic forces which could be responsible for the large k-values.

In 1994 it was finally realized that figure 6.6 may represent a special case in which there existed an additional anomalous force which was proportional to the Ampère force. The following experiments with very efficient energy conversions revealed, however, clear signs of the variation of the experimentally determined k-values with shot conditions, other than the arc geometry.

Since every force of nature is coupled to its own particular store of energy, it became inevitable that the anomalous water plasma forces had to involve some energy which was not stored in the capacitors and yet gave rise to some of the explosion force. Then quite naturally the question has to be asked: what is the origin of the additional energy? Finding the answer to this question became the primary motivation of research in three laboratories located in Toronto, Canada, Richmond, Virginia and Oxford, England.

The magnitude of the anomalous force, in relation to the electrodynamic force, can be probed by finding an experimental value of the dimensionless k-constant and denoting it by k'. This k' is then given by the measured momentum mv_0, acquired by a projectile accelerated by the explosion, and the action integral A of the current pulse. The analysis of the general arc behavior outlined in Chapter 6 is relevant to this problem. According to Eqs.6.13 and 6.14, the action integral is a function of the current intercept I_0 of Eq.6.8 at t = 0 and the time constant T of the ringing capacitor discharge current. Recognizing that the initial velocity v_0 of the projectile is given by Eq.6.5, it is clear that the empirical k' can be deduced from the experimental measurements of the projectile mass m, the throw height h, the current intercept Io and the time constant T. It may be written

$$k' = \frac{4\pi}{\mu_0} \frac{4m\sqrt{2gh}}{I_0^2 T} \qquad (7.10)$$

A series of measurements were carried out by Hathaway Consulting Services (HCS) of Toronto to test Eq.7.10. The Toronto accelerator was not the cartridge of figure 6.5, but a device similar to figure 6.8. Figure 7.15 depicts the Hathaway accelerator. It relied on a Formula 1 racing car spark plug as the breech electrode. The purpose of this was to employ only inorganic materials to contain the explosion. With the ceramic and metal components the explosion cavity was far more rigid than the fiber-glass reinforced epoxy cartridge. The rigidity of the cavity walls determined how much strain energy was absorbed and subtracted from any propulsion mechanism. Therefore, with the all-ceramic-and-metal accelerator of figure 7.15, the hope and anticipation was that a greater amount of anomalous energy would be passed on to the projectile mass resting on the water.

As figure 7.15 reveals, the water volume v was contained in a thick steel barrel and between the spark plug and a steel piston, M, resting directly on the water. The mass of the piston was large compared to the mass of the water, so that the piston would only be lifted by a few millimeters when the energy discharged from the capacitor varied between 25 and 57 J. The throw-height, h, was measured with a precision dial gauge. Typical discharge current oscillograms are shown in figure 7.16. The first two peaks were used to evaluate I_0 and T. The capacitor was charged to voltages V_0 ranging from 10 to 12 kV. The piston mass, M, varied slightly between 89.6 and 101.8 gm due to small changes in attachments required for the throw-height measurement of 1 - 5 mm.

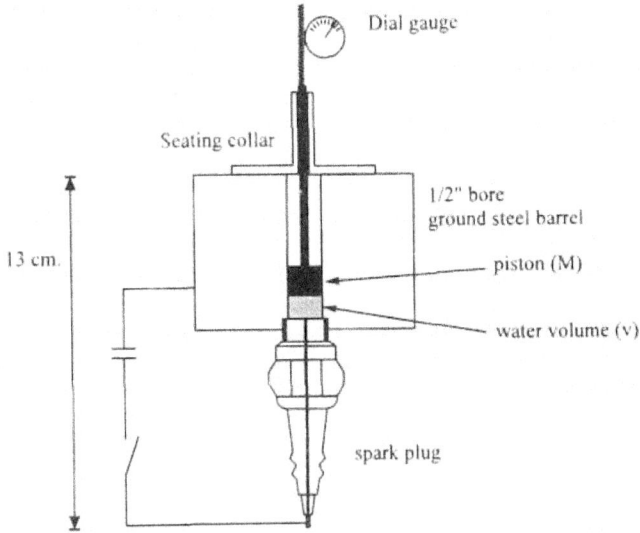

Figure 7.15 : Simplified diagram of the Hathaway Plasma Accelerator

$V_0 = 10 \, kV, \; C = 0.5 \, \mu F, \; L = 0.99 \, \mu H$

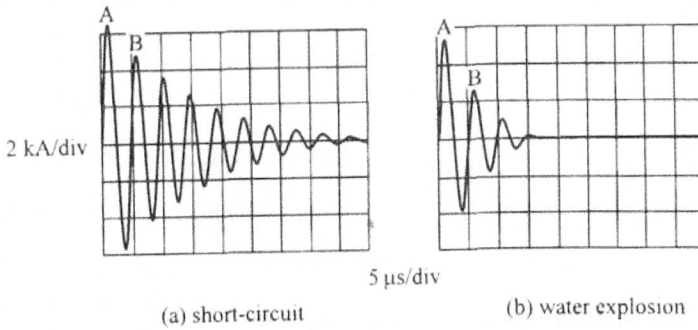

(a) short-circuit   (b) water explosion

Figure 7.16 Typical current oscillograms measured using the accelerator of figure 7.15

A common factor in all the shots was the 0.5 μF capacitor. The water charge was 1.5 ml of distilled light water (H_2O). The results of 14 shots are listed in table 7.2 in ascending order of voltage. This table also gives the computed values of the action integral A from Eq.6.14 as well as the time-average lift force <F>. The same results will be used later to analyze the stored pressure energy. In the average force calculation the pulse duration was taken to be that given by the time constant T of the decaying, ringing current pulse.

Figure 7.17 is a plot of the mechanical impulse ($\int F\, dt$) received by the piston of figure 7.15 versus the action integral ($\int i^2\, dt$) of the current pulse for the results listed in table 7.2. This graph should be compared with figure 6.6 for the saltwater cartridge. The less rigid plastic cartridge gave a straight line as if the process involved only electrodynamic energy. In the more effective Hathaway plasma accelerator there is no proportionality of impulse to action integral. Therefore the prominent explosion forces in the latest plasma accelerator cannot be electrodynamic forces. They must therefore be anomalous forces of non-electrodynamic origin. This is a very remarkable finding which is not only of great scientific interest but could have major technological consequences. There is now little doubt that the water gives up considerable amounts of internal energy which is not easily liberated by any other method.

Shot #	V_0 (kV)	M (gm)	h (mm)	I_0 (kA)	T (μs)	A ($A^2 s$)	<F> (N)	k'
POLD12	10	102	2.97	6.27	9.64	94.7	2550	2600
P12301	10	90.6	2.60	5.83	6.50	55.2	3150	3700
P12519	10	90.6	2.18	6.43	6.78	70.1	2760	2670
P12718	10	89.6	3.61	5.85	5.44	46.5	4380	5120
P12722	10	89.6	3.18	5.81	6.26	52.8	3570	4230
P12302	11	90.6	2.67	6.53	7.25	77.3	2860	2680
P12521	11	90.6	2.54	7.15	8.34	107	2420	1900
P12720	11	89.6	3.71	6.33	6.49	65.0	3720	3720
P12727	11	89.6	3.20	6.47	7.75	81.1	2900	2770
P12405	12	90.6	1.75	6.75	10.2	116	1650	1440
P12407	12	90.6	4.52	6.76	8.44	96.4	3200	2800
P12717	12	89.6	3.81	6.98	7.22	87.9	3390	2780
P12726	12	89.6	3.71	6.63	9.83	108	2460	2240
P12403	12	90.6	2.90	7.35	9.24	125	2340	1730

Table 7.2 : Results of mechanical impulse measurements using the accelerator of figure 7.15

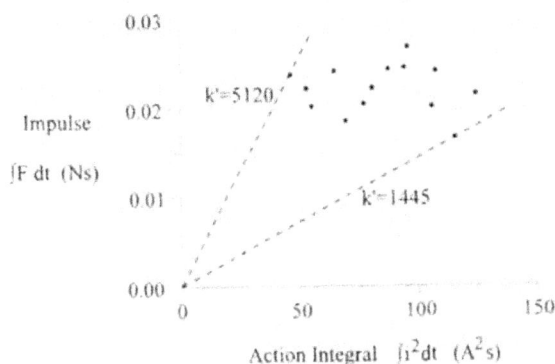

Figure 7.17 Mechanical impulse versus action integral for the results shown in table 7.2

2.   Source of the Anomalous Energy

Energy conservation is a vital aspect of the Newtonian electrodynamics. Therefore it is disconcerting to find energy without an identifiable source. Experiment has shown that the anomalous energy is not of thermal or electrodynamic origin. Where could it come from?

The holy grail of modern physics is quantum field theory, as first developed by Dirac [7.20]. He discovered that, according to this theory, the vacuum should contain negative kinetic energy. This was subsequently identified with the energy of positrons (positive electrons). It is now generally believed that electrons and positrons continuously annihilate each other in the vacuum with the emission of electromagnetic energy (photons), only to reappear later and elsewhere, in particle pair creation which absorbs energy. Quantum field theory, therefore, lodges large amounts of energy in the vacuum. Perhaps some day we may be able to tap this energy for useful purposes.

The vacuum energy is said to reveal itself in the motion of matter particles at absolute zero temperature. For this reason it is also called 'zero point fluctuation'. Puthoff [7.21], one of the principal exponents of zero point fluctuations, believes that they provide energy to the electron of the hydrogen atom to replenish the energy radiated from the atom by the orbiting electron. By this means he ensures stable energy levels in the atom which is the primary objective of quantum theory. Puthoff argues that the Casimir attraction of two closely spaced metal plates is the result of zero point fluctuations and could become the vehicle of gaining energy from the vacuum.

The effect of zero point fluctuations is a form of radiation pressure. It is difficult to see how this could manifest itself in water plasma explosions and not in air plasma explosions. It has been shown in Chapter 6 that the latter are driven by the electrodynamic Ampère force alone. Water plasma explosions require an additional force which arises from the nature of water and not from vacuum, the underlying vacuum space being common to both water and air plasmas. Hence we have to explore other sources of energy which are related to matter.

The nuclear energy liberated in filament fusion experiments, and possibly some cold fusion experiments, comes to mind. As outlined by Storms [7.3], there is little doubt that excess heat is generated in certain electrolytic experiments with metal electrodes and heavy or light water. Plasmas in light and heavy water explode with almost equal force. Heavy water seems to provide a little more force, as will be seen from some following experimental results, but the difference is only ten percent or less. This is too small to indicate that nuclear reactions play a significant role. While filament fusion experiments have generated a copious amount of easily detectable X-radiation, no X-rays have been found to accompany water plasma explosions. X-ray films have been placed directly over the muzzle of water plasma accelerators. The film was mechanically protected from water impact by a transparent barrier layer, but no trace of an X-ray exposure was found. This seems to rule out nuclear energy as the possible source of the anomalous energy.

The difference between water vapor and liquid water is the force of attraction which binds the molecules of the liquid together. Like all other forces of nature, these attractions have to be the result of stored energy which, in this case, may be called the energy of liquid cohesion. Just as with gravitation, attractions involve potential energy. This is energy which depends on the position of, and the distance between, the interacting matter particles. It is not motion dependent kinetic energy.

Bonding forces between water molecules, which provide for liquid cohesion, must not be confused with the atomic bonding forces between hydrogen and oxygen inside the H_2O molecule. The latter bonding forces are far stronger than the intermolecular forces and, consequently, involve a greater quantity of bonding energy. Bonding energy between atoms is normally described as chemical energy. Since the microscopic forces driving plasma explosions are unable to create steam by breaking all of the intermolecular bonds, they are certainly too weak to dissociate hydrogen from oxygen.

If it were possible to regain some of the potential energy invested in the liquid during the condensation process, how would this liberated potential energy manifest itself ? Here it should be interjected that modern physics accepts the universal validity of Coulomb's law, yet it has no straight-forward answer to the question of what counteracts the attraction between the electrons and the nucleus of an atom. The whole machinery of quantum theory was invented to find arguments for stable energy levels in which the electron may be situated or orbit, yet the repulsive reaction force which must support this stability remains shrouded in mystery. The energy levels have been accepted as empirical facts somehow supported by field theory. Starting with Newton's physics, it could be argued with equal conviction that the reaction forces of repulsion between molecules, in equilibrium with the bonding forces of attraction, are also an empirical fact.

Treating the repulsive reaction forces as empirically given, a reduction in bonding force might not instantly result in a reduction of the repulsive reaction force. The repulsions would then lead to a separation of the matter particles until the falling force of repulsion, governed by particle distance, is once more in equilibrium with the new force of attraction. This argument applies equally to interatomic and intermolecular forces which are all subject to quantum mechanical energy levels. Therefore a sudden reduction in bonding forces between liquid water molecules, and the accompanying liberation of potential energy, could manifest itself in an explosion, just as observed in the water plasma experiments.

To explore this idea further we have to investigate the bonding forces between molecules in liquid water, an area still full of unanswered questions. In a recent paper in Nature [7.22], dealing with the molecular structure of water, we find the following statement:

> "Liquid water, the medium in which life both began and persists, is in many ways a most unusual fluid. Much is known about the macroscopic properties of the condensed and gaseous states of water, but our understanding of the microscopic forces that define water structure remains incomplete."

The water molecule is highly unsymmetrical. The two small hydrogen atoms are not attached to opposite sides of the large oxygen atom, but clustered on one side. The result is a strongly dipolar molecule. For this reason the force of attraction between two water molecules can have a variety of values, depending on the dipole orientations. Hydrogen atoms on one molecule seem to prefer bonds to hydrogen atoms on other molecules, rather than attach themselves to oxygen atoms. The H-H bonds between molecules are particularly strong, and are called 'hydrogen bonds'. Potential energy may well be liberated when the hydrogen bonds are broken or distorted.

In a liquid the molecules are in continuous motion. Their polar orientations could be in complete disorder. A certain order has, however, been revealed by X-ray diffraction studies. Molecular order is also inferred from the measurement of magnetic susceptibility and the dielectric constant. Locally this order may be changing all the time, but it seems that some global average order is being maintained. The observed structure is not too far removed from that of ice crystals. Tetrahedral structure is most frequently mentioned [7.23, 7.24] in which the oxygen atoms are placed at the corners of a tetrahedron.

Dorsey [7.25] distinguishes between the linkage of water molecules and the architecture of water structure. He argues that in a collision of two molecules in the liquid, because of their dipolar nature, the molecules remain for longer or shorter times close together and are mutually oriented in a preferred manner. The architecture of the water structure must be due to a longer range of order of typical associated groups of molecules. Both of Dorsey's mechanisms contribute potential bonding energy.

When water is evaporated, the bonding forces of whatever structure there might exist, are overcome by thermal motion of the molecules. Thermal motion is dependent on the supply of heat energy to the liquid. The latent heat of evaporation, as it is called, amounts to 540 calories per gram of water at the boiling temperature. This means the molecular bonding energy is 2260 J per ml of water at normal pressure and temperature. It is the maximum potential energy that could be extracted from ordinary water. Considering the total amount of water on earth, it represents an enormously large reservoir of renewable energy. It may be described as 'renewable' because all the potential bonding energy is recovered when water vapor condenses and stores energy which it acquired from the atmosphere.

Given the uncertainty of liquid water structure, it is entirely possible that the bonding energy in a small fog droplet is less than 540 cal/gm. Should this prove to be the case, then the difference in the potential energies would be liberated explosively when liquid water is suddenly transformed to fog. This behavior is observed in water plasma explosions.

Figure 6.16 depicts the explosive transformation of water to fog. R. Hull [7.26] of the TCBOR laboratory in Richmond, VA, captured the fog plume from 1.5 ml of water volume with a video camera exposure of one millisecond. He discharged 144 J of energy from a capacitor into the plasma accelerator. The conical tip of the plume indicates that the fog was travelling at supersonic speed and thereby created the characteristic Mach conical shockfront. With lesser amounts of energy in the capacitor the fog jet formed a subsonic mushroom cloud.

After the tip of the jet struck the laboratory ceiling, the fog rolled around like a cloud and evaporated. Fog droplets have to be less than 100 μm in diameter, otherwise they will fall like rain. It is the bombardment of fog droplets by air molecules which keeps clouds aloft in the sky. The same bombardment also suspends heavier-than-air dust particles in the atmosphere.

As will be more fully revealed later, less than half of the 1.5 ml of water was converted to fog. The remaining water in a mixture with air followed the fog out of the accelerator at much slower speed. The water then fell back on the accelerator and formed quite large drops on metallic surfaces.

The atomic and molecular structures of some solid particles of matter are very different from the structures found in the bulk materials. The best known example is the giant carbon molecule C_{60}, described as Buckminsterfullerene, and discovered as recently as 1985 [7.27]. Its structure takes the form of a Buckminster Fuller geodesic sphere consisting of 12 pentagons and 20 hexagons with the symmetry of a soccer ball. It represents the third allotropic form of carbon in addition to the diamond crystal and amorphous graphite. The fullerene has properties which differ greatly from the other two carbon substances, all because of its own individualistic potential bonding energy. The discovery of fullerenes has given great impetus to the physics and chemistry of small clusters of matter. A fog droplet is similarly a small cluster of molecules.

No mention has been made, so far, of the surface tension of liquid water. A molecule in the water surface has many neighbors on one side and none on the other. Normal forces of attractive cohesion will, therefore, pull the surface molecule into the liquid. The surface tension energy is bonding energy which forms part of the total potential energy stored in the liquid. When a drop of water is divided into two drops, additional surface area is being created. This requires additional surface tension energy which is the mechanical work that has to be done in pulling the two drops apart. The dividing motion is resisted by something like tension in the water surface. This has given rise to the term 'surface tension'.

When water is divided into fog droplets, work has to be done in separating the droplets. This could be achieved by the electrodynamic Ampère forces of repulsion present in the water. The work done by these forces would be stored in the fog as additional potential energy of surface tension and could be regained when molecular bonds inside the droplets are broken. This would be the link between electrodynamics and water plasma explosions. A link must exist, because no explosion takes place without current flow.

The ordinary intermolecular bonding energy, besides that of surface tension, must obey the rules of quantum mechanics. As in all matter aggregates, only certain arrangements will be stable and they are defined by quantum mechanical energy levels. Hence the transformation of one particular structure pattern to another pattern is the result of a quantum leap in bonding energy. In this way it becomes feasible that a small investment in surface tension energy, which makes fog, could trigger a quantum leap of bonding energy release which is greater than the added surface tension energy. This furnishes a complete qualitative hypothesis for the mechanism of the liberation of anomalous (bonding) energy in water plasmas subjected to a strong current pulse. Without the electric current however, the potential energy remains locked in the liquid water structure.

3. Anomalous Pressure Energy

In 1986 we reported pressure measurements of powerful saltwater plasma explosions [6.7]. The observed pressures, in excess of 20,000 atm, could not be explained with Joule heating nor the electrodynamic forces of field theory or the Newtonian electrodynamics. An idea of the size of the anomaly can be gained from the general electrodynamic force law, Eq.6.15, in which k is the dimensionless constant depending on circuit geometry and the electrodynamic theory employed.

The only way in which the Lorentz force of relativistic electromagnetism could account for plasma pressure is via magnetic pinch action. Northrup [2.8] has shown that this results in a k-value of 0.5, whereas measurements gave k>6000. Depending on circuit geometry, Ampère's force law allows for 10<k<200. This is more realistic than the prediction of the Lorentz force, but still falls far short of observations. From these facts it has to be concluded that the forces developed in water plasma explosions are at least 90 percent anomalous. It seems inevitable that they must also involve anomalous energy, although not in the same proportion.

The exponentially decaying current oscillation of a capacitor discharge, as described by Eq.6.8, theoretically lasts forever, however the action integral of Eq.6.12 has been shown in approximation, Eq.6.14, to have finite duration, and that a good practical measure of it is the time constant T. So far, experiments have provided no evidence of the pressure pulse lasting longer than the associated current pulse. For example, no loss of water acceleration was noticed when the length of the accelerator barrel, from breech to muzzle, was reduced from 10 cm to 2 cm. Present indications are that a water column of 1 cm length expands no more than 0.5 cm during the acceleration period of a low energy pulse.

There is clearly a connection between current flow and pressure generation. No pressure can be produced without current. Faced with this fact, and until contrary evidence is brought to light, we have little choice but to assume that the duration of the pressure pulse is equal to the duration of the current pulse. The time constant T is then a reasonable measure of the pulse duration. In this way we can define an average explosion force <F> by

$$\int F \, dt \ = \ <F> \, T \tag{7.11}$$

This is not the same average force which was defined by Eq.6.17, which represented the average electrodynamic force, but is instead the experimentally measured average force.

It is the nature of hydrostatics to convert forces in water to hydrostatic pressure acting in all directions. Let the piston area be S, which in figure 7.15 is the pressure surface causing the impulse ($\int F dt$) on mass M. The time average increase in pressure above atmospheric pressure may then be written

$$<\Delta p> = \frac{<F>}{S} \qquad (7.12)$$

The volume contributing to the force and pressure generation is the water volume v. Noting the equality of impulse and momentum of Eq.6.6, and using Eqs.6.7 and 7.12, the time average pressure energy stored in the plasma cavity can be expressed as

$$<\Delta p> v = \frac{M v \sqrt{2 g h}}{S T} \qquad (7.13)$$

where h is the height to which the mass M is thrown, and g is the acceleration due to gravity.

Current oscillograms of the capacitor discharges were recorded. Figure 7.16 shows typical traces. Trace (a) is for a test in which the water accelerator was short-circuited, and (b) depicts the trace for a typical discharge through water. It will be noted that the water caused severe damping of the current. Part of this must have been due to Joule heating of the water. The remainder must have been the result of mechanical work performed by the electrodynamic forces in straining the explosion cavity and, to a lesser extent, lifting the piston. Doing mechanical work by the arc (motor effect) causes the induction of a back-e.m.f. which is in phase with the resistive voltage drop and indistinguishable from it in measurements.

Without the plasma accelerator, the circuit behaved like any ordinary LCR circuit. Even with water explosions this behavior persisted as far as the influence of inductance on ringing frequency and capacitance on surge impedance were concerned. Induced back-e.m.f., however will have added to the damping produced by ohmic heating.

Denoting the first positive current peak of an oscillogram (see figure 7.16) by A and the second positive peak by B allows the calculation of I_0.

$$I_0 = A \exp\left(\frac{\ln (A/B)}{4} \right) \qquad (7.14)$$

From the ringing frequency f, the time constant is given by

$$T = \frac{1}{f \ln (A/B)} \qquad (7.15)$$

It is of interest to know how much of the capacitor energy is wasted in the external circuit of the accelerator. This can be determined with a test in which the accelerator electrodes are short-circuited. The result is a trace like (a) of figure 7.16. If R_C is the resistance of the circuit when short-circuited, and the stored capacitor energy, E_S, is discharged, we have

$$E_S = R_C \int i^2 dt \approx \frac{R_C I_{0C}^2 T_C}{4} \qquad (7.16)$$

where I_{0C} and T_C are derived from the short-circuit oscillogram using Eqs.7.14 and 7.15. The value of R_C obtained from Eq.7.16 may then be used in plasma explosion calculations. For the appropriate I_0 and T of the explosion, the energy lost in the circuit, E_C, is

$$E_C = \frac{R_C I_0^2 T}{4} \qquad (7.17)$$

This energy loss is not available for generating pressure and driving the explosion.

Furthermore, the current oscillogram of a plasma explosion allows an estimate to be made of the ionization energy. Let the ionization loss reduce the initial capacitor voltage by ΔV before the current reaches its full value. The energy remaining in the capacitor after ionization is then

$$\tfrac{1}{2} C (V_0 - \Delta V)^2 = \tfrac{1}{2} C (V_0^2 - 2 V_0 \Delta V + (\Delta V)^2) \qquad (7.18)$$

and thus the ionization energy E_I can be written

$$E_I = C V_0 \Delta V - \tfrac{1}{2} C (\Delta V)^2 \qquad (7.19)$$

The total impedance of the circuit is the sum of the resistance R_C and the surge impedance, Z_0, of Eq.6.10, with L being the circuit inductance given by

$$\omega = 2\pi f = \frac{1}{\sqrt{LC}} \qquad (7.20)$$

The circuit parameters in the pressure energy measurements performed in the HCS laboratory in Toronto, to which table 7.2 refers, were

$R_C = 135\ m\Omega$
$L = 0.99\ \mu H$
$C = 0.5\ \mu F$
$Z_0 = 1.41\ \Omega$.

Hence ΔV could be calculated from

$$\Delta V = V_0 - I_0(Z_0 + R_C) \qquad (7.21)$$

and then substituted into Eq.7.19. The ionization energy, E_I, does not generate pressure in the plasma cavity and, together with the circuit loss, E_C, it should be subtracted from E_S to arrive at the net electrical energy available for pressure generation.

Figure 7.18 presents pressure energy results calculated with Eq.7.13 and the results in table 7.2 for three voltage levels. The surface area of the piston, S, was the cross-section of the ½" cylindrical bore of the vessel or 1.27×10^{-4} m², and the volume of the explosion cavity, v, was 1 ml. The combined circuit and ionization loss was computed with Eqs.7.17 and 7.19. A certain variability of the loss figures was probably due to differences in the electrolytic conduction loss during the ionization phase. On the whole, the average loss increased from 47 percent of the stored capacitor energy at 10 kV, to 50 percent at 11 kV, and to 61 percent at 12 kV. It must be stressed that these loss figures do not include an allowance for the Joule heat generated in the plasma and are therefore conservative. If maximum efficiency of the energy conversion process is the objective, then attention will have to be paid to the loss mechanisms. In the reported investigation no effort was made to keep R_C small. In fact the external circuit included a switching arc in air which may have wasted more energy than the metallic circuit.

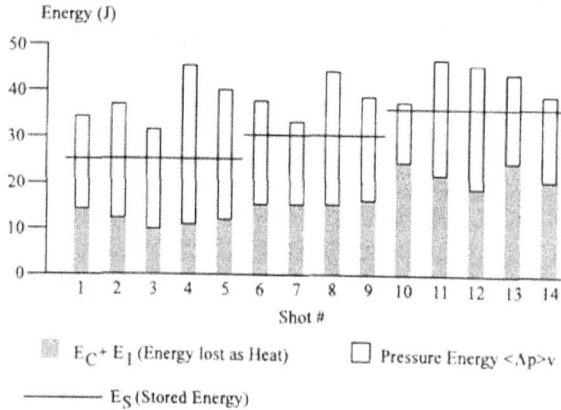

Figure 7.18 : Anomalous pressure energy for the 14 shots described in table 7.2

The 'available' energy of figure 7.18 is the difference between the stored input energy E_S and the energy lost as heat (E_C+E_I). This difference is the energy actually present in the arc cavity to generate pressure and cause anomalous energy to be liberated. In all of the shots shown in figure 7.18, the sum of the energy lost as heat combined with the time-average pressure energy $<\Delta p>v$ from Eq.7.13, is greater than the energy initially stored in the capacitor. Thus in each case the amount of energy above the stored energy line, E_S, is anomalous. If the energy conversion efficiency is defined by a gain factor Q_p, so that

$$Q_P = \frac{\text{measured pressure energy}}{\text{available input energy}} = \frac{<\Delta p> v}{E_s - E_c - E_l} \qquad (7.22)$$

we find from the results of figure 7.18 that Q_p varies between 1.13 and 2.44. No explanation of this wide range of gain factors has emerged. Statistical fluctuations in the dielectric breakdown process introduce variability into all arc experiments, however the variation in the liberation of anomalous energy seems to go beyond this fluctuation and could well be inherent in the energy transformation process. Overall, there is good reason to conclude that significant amounts of energy from light water may be tapped for technological ends.

Another important result of the pressure energy investigation is presented in figure 7.19. This compares heavy water (D_2O) with light water (H_2O) and a saturated solution of saltwater ($H_2O+NaCl$). The saltwater cup experiments of Chapter 6 proved that electrolytic currents do not exhibit the same magnetic action as conduction currents. Saltwater is a good electrolytic conductor and, until it breaks down and conducts as a plasma, no explosion forces are developed. This is the reason for the small impulses generated with saltwater, as indicated in figure 7.19.

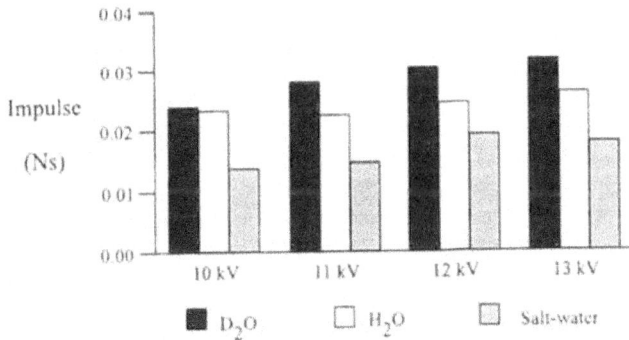

Figure 7.19 Comparison of mechanical impulse generated using different liquids

One of the saltwater shots provided good quantitative information on the electrolytic action. The 0.5 µF capacitor was charged to 12 kV. Using Eq.7.21, ΔV was nearly 5 kV. Thus by Eq.7.19, it was found that 23.8 J of the 36 J stored in the capacitor were used to drive Na^+ ions to the cathode and Cl^- ions to the anode and ionize the water. Since ionization of distilled water has been previously shown to require about 5 J, it appears that the electrolytic conduction process in saltwater absorbed 19 J or 53 percent of the capacitor energy.

In this particular shot the circuit loss was 3.7 J which, in addition to the 23.8 J of electrolytic conduction and ionization loss, left 8.5 J for generating pressure and liberating anomalous energy. This was done quite efficiently by producing 35.8 J of pressure energy. It was, however, still less than the input energy and fared badly in the comparison of figure 7.19.

The comparative tests were performed to decide if heavy water would contribute nuclear energy. As revealed by figure 7.19, D_2O furnished greater explosion impulses than H_2O, but the advantage was small, of the order of ten percent. Nuclear energy would be expected to produce more spectacular results. Under the circumstances, therefore, it seems safer to speculate that internal water energy (potential energy of molecular bonding) is responsible for the difference in the measured impulses. This experiment demonstrates that the magnitude of the anomalous plasma explosion force does depend on the composition of the water plasma.

As there was obviously no excess pressure left in the explosion cavity after the current pulse, the pressure energy must have eventually been converted to heat energy. With a light projectile a significant fraction of the pressure energy could probably be converted to kinetic energy.

4.   Anomalous Kinetic Energy

For transient pressure energy to become useful, this energy has to be converted to heat, or mechanical work as in heat engines, or to kinetic energy of moving objects. Of the three alternatives, kinetic energy is the most attractive because it may be used to drive electricity generators without incurring the penalty of the Carnot efficiency limit of heat engines. That considerable amounts of kinetic energy can be achieved with water plasma explosions was demonstrated at MIT [6.9] with the device of figure 6.8 which punched a hole through an aluminum plate, as illustrated by figure 6.9.

Assuming that all of the anomalous energy that is liberated in the explosions is potential energy of liquid cohesion, and that whatever electrodynamic energy was available in the cavity was absorbed as surface tension energy in the fog formation, the first result of the process is to create pressure energy in the explosion cavity. This is indicated by the energy flow chart of figure 7.20. Concurrent with the rise of pressure, some of the associated energy is immediately shunted into the cavity walls where it will stress elastic materials and do mechanical work on inelastic ones by plastic deformation and destruction of atomic and molecular bonds. The only way of limiting the loss of deformation energy is to provide the strongest and most rigid explosion cavity. With infinite rigidity the deformation energy would be zero.

Pressure in the water is likely to cause molecular collisions in the water and at the walls. Molecular motions will thereby be randomized and this form of kinetic energy is heat. In an infinitely rigid enclosure all of the pressure energy will be converted to heat. This process is expected to be fast for, at this stage, no water pressure has been observed to persist after the current has ceased to flow.

If water can escape from the cavity, or one of the cavity walls is mobile, then the pressure will accelerate matter and generate kinetic energy. Experimental circumstances are likely to determine how much pressure energy can be converted to kinetic energy, and how much will escape as useless low-grade heat. The present phase of research on water plasma explosions deals with this energy conversion mechanism. All results obtained so far are preliminary and may be upset by further research.

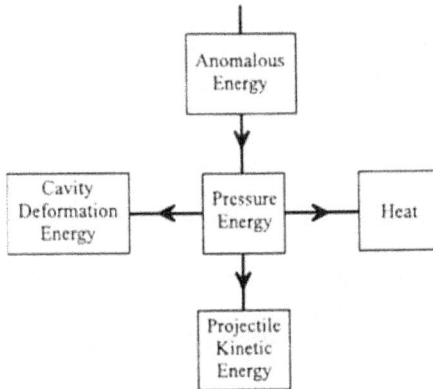

Figure 7.20 : Energy flow chart

A projectile accelerated by the explosion of a fluid receives a mechanical impulse ($\int F\, dt$). If the projectile is unrestrained, Newton's laws of motion require that the impulse be converted to projectile momentum in accordance with Eq.6.6. Given the impulse, and denoting the initial projectile velocity by u_0, the kinetic energy of the projectile may be expressed as

$$E_K = \tfrac{1}{2} m u_0^2 = \tfrac{1}{2} u_0 \int F\, dt \tag{7.23}$$

For maximum kinetic energy, therefore, the initial velocity should be a maximum. From Eq.6.6 this velocity is

$$u_0 = \frac{\int F\, dt}{m} \tag{7.24}$$

Substituting Eq.7.24 into Eq.7.23 leads to

$$E_K = \frac{(\int F\, dt)^2}{2\, m} \tag{7.25}$$

Hence reducing the mass to one-half its original value doubles the kinetic energy for the same impulse. To convert pressure energy most effectively to kinetic energy, Eq.7.25 indicates that the smallest possible projectile mass should be chosen.

There are practical limits to this method of maximizing the kinetic energy. The mass cannot be reduced below the water mass needed to produce the full force of the explosion. When the water volume in the accelerator becomes too small it may be difficult to produce sufficient impulse, and the kinetic energy is proportional to the square of this impulse. If a solid projectile is placed on the water, the total mass to be accelerated increases and this will lower the measured kinetic energy.

Early in the search for kinetic energy the collaborating TCBOR Laboratory of Richmond, VA, proved the conversion of water to fog as demonstrated by video photographs like figure 6.16. The accelerator used in Richmond also resembled the miniature coaxial water arc gun of figure 6.8. It comprised a half-inch bore steel barrel of 10 cm length from breech to muzzle. The breech electrode was a 3/16-inch diameter bronze rod insulated with a Nylon sleeve. Nylon was also used as the secondary insulation between the barrel and the copper base plate. The explosive transformation of 1.5 ml of distilled water to the supersonic fog jet of figure 6.16 was photographed with an exposure time of one millisecond. A drizzle of small droplets rained back from the drifting cloud of fog which evaporated in a matter of seconds.

With lesser amounts of energy in the capacitor, the fog jet was a subsonic mushroom cloud. This was confirmed in the Department of Engineering Science of Oxford University with a high-speed framing camera. The development of the mushroom cloud is shown in the photograph of figure 7.21. With sufficient energy this type of fog jet will also strike the ceiling and cause a blue flash on impact. This is visible with the naked eye because of the sudden concentration of luminous ions. The ion concentration in the jet can also be seen in the first few centimeters above the muzzle, but then it decreases below the visibility level until the ions are stopped and diverted laterally. For a few seconds the fog rolls around under the ceiling until it all evaporates.

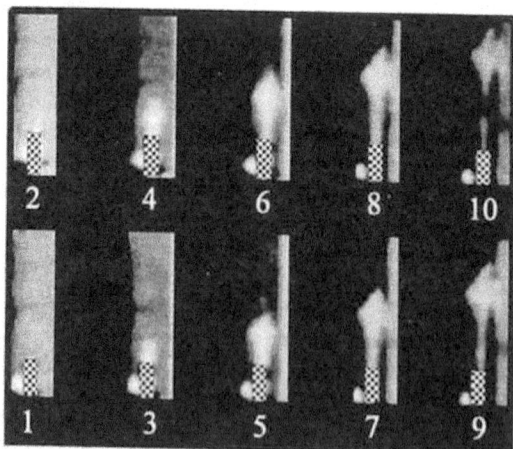

Figure 7.21 High speed photography of the development of a mushroom cloud above the accelerator (10,000 frames per second) Gun barrel has been highlighted with checker pattern for clarity

High speed photography allowed the tip speed of the fog jet to be measured, but neither the mass of the fog travelling at this speed nor the velocity distribution was known. To overcome this problem a secondary projectile technique was developed and tested in the Toronto laboratory. This relies on the principle of momentum conservation, and is largely unaffected by air motion and ionization because of the negligibly small air mass involved.

A secondary projectile of mass M is placed on the muzzle of the plasma accelerator. It should be made of a material into which fog can be driven and absorbed. Balsa wood was found to be suitable for this purpose. If the mass, M, absorbs a mass, m, of water, and the two masses travel together at the initial velocity v_0, momentum conservation predicts

$$m\,u_{av} = (M+m)\,v_0 \qquad (7.26)$$

where u_{av} is the average velocity of the water. The throw height, h, of the secondary projectile was usually less than 1 meter, and was measured with a video camera. Energy conservation demanded that

$$\tfrac{1}{2}(M+m)\,v_0^2 = (M+m)\,g\,h \qquad (7.27)$$

or

$$v_0 = \sqrt{2gh} \qquad (7.28)$$

where g is the acceleration due to gravity. The average velocity may then be calculated with

$$u_{av} = \frac{M+m}{m}\sqrt{2gh} \qquad (7.29)$$

This average velocity gives the minimum kinetic energy of the absorbed water to be

$$E_{K,min} = \tfrac{1}{2}m\,u_{av}^2 \qquad (7.30)$$

The true kinetic energy of the water depends on the root mean square (r.m.s.) velocity of the fog jet because for n water particles of mass Δm travelling at the velocity u

$$E_K = \sum_n \tfrac{1}{2}\Delta m\,u^2 = \tfrac{1}{2}u_{rms}^2\sum_n \Delta m = \tfrac{1}{2}m\,u_{rms}^2 \qquad (7.31)$$

Kinetic energy measurements were carried out with a spark plug accelerator which was similar to figure 7.15. Two different barrel lengths (breech to muzzle) of 5 cm and 2 cm were evaluated. Modified racing car spark plugs survived two to five shots depending on the magnitude of the explosion forces. The unreliability of the spark plugs is believed to be one of the causes of the variability of the results.

Figure 7.22 depicts the two secondary projectiles employed in the investigation. R denotes a rocket-shaped steel can with nose cone and a partial filling of balsa wood. The soft wood stood directly on the muzzle of the accelerator. The overhanging steel skirt was meant to restrain the projectile to vertical flight. R tumbled in the air in spite of the skirt. In fact the skirt seemed to have a detrimental effect on the throw height. Experiments indicated that the metal can and the nose cone were not really needed. A simple balsa wood cylinder weighed down with steel washers was found to perform better, although it tumbled even more. It was therefore called a tumbler and denoted by T. Because of the tumbling, the measurements of the throw height h had to be based on the rise of the center of gravity of the secondary projectiles. Rotational kinetic energy was ignored.

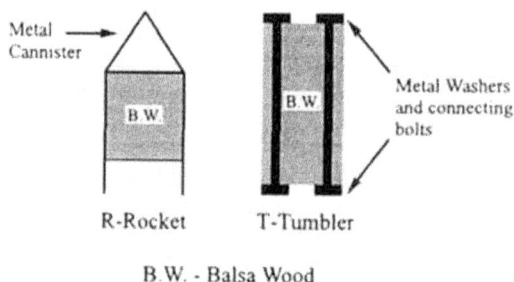

Metal Cannister

B.W.

Metal Washers and connecting bolts

R-Rocket   T-Tumbler

B.W. - Balsa Wood

Figure 7.22 : Two types of secondary projectile containing balsa wood

The discharge circuit used by HCS of Toronto comprised a $C = 0.565 \ \mu F$ capacitor, a triggered air spark gap and a current transformer. A feature of the circuit was its low inductance of $L = 0.31 \ \mu H$. This resulted in a surge impedance of $Z_0 = 0.74 \ \Omega$. With $V_0 = 12$ kV, the most frequently used charging voltage, the circuit was capable of producing more than 15 kA maximum current with a time constant of the order of 2 μs. This produced much stronger electrodynamic forces than had been experienced in the pressure energy measurement program. With the very low inductance, spark plugs survived on average no more than two discharges. Therefore, to obtain less violent explosions, a conductor loop was inserted in the circuit. This raised the inductance to $1.24 \ \mu H$.

To determine how much of the kinetic energy of the flying water mass was due to the liberation of internal water energy, it was necessary to measure energy losses and arrive at a figure of the energy supplied to the water which was available for accelerating water. Only in the case of the high inductance circuit did it prove possible to derive heat losses from the current oscillogram.

A sample loss calculation will now be outlined in conjunction with figure 7.23. This is the oscillogram of a test in which the accelerator was short-circuited. It resulted in a slight reduction of the inductance to $L = 1.18 \ \mu H$. The open gap of the switching arc interrupted the tail of the discharge current, leaving a residual voltage V_r on the capacitor. For the charging voltage V_0, this made the stored capacitor energy, E_S, which was actually dissipated in the circuit

$$E_s = \tfrac{1}{2} C (V_0^2 - V_r^2) \qquad (7.32)$$

In the short circuit test virtually all of E_S was converted to heat by (1) ohmic heating of the metallic circuit and the switching arc plasma, and (2) ionization of the air arc in the switch. The energy of (2) is recovered as heat on recombination of the ions. The ohmic heat loss will be denoted by E_C and the ionization energy by E_I. If the first and second negative peaks of the underdamped current of figure 7.23, represented by Eq.6.8, are labelled A and B, I_0 may be computed with Eq.7.14. In the case of figure 7.23, this came to $I_0 = -5.91$ kA. The oscillogram indicates a frequency of $f = 195$ kHz, and using Eq.7.15, the time constant comes to $T = 34.9$ μs. When all of the current damping is the result of resistive losses, the circuit resistance R_C may be calculated with Eq.6.11, and is found to be $R_C = 68$ mΩ.

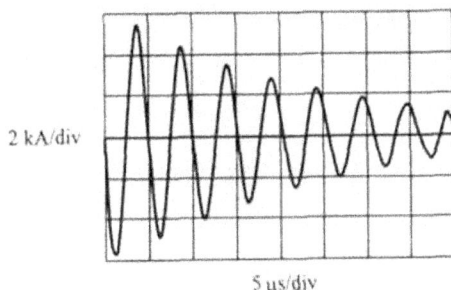

2 kA/div

5 μs/div

Figure 7.23 Current oscillogram for short-circuit across accelerator

The total Joule (ohmic) heat is given by Eq.7.17. For the case considered it became $E_C = 20.7$ J. The energy supplied to the circuit, from Eq.7.32, was 39.8 J. The difference between these two figures should be the ionization energy of the air in the switch. It is, therefore, expected to be 19.1 J. Given that $V_0 = 12$ kV, this last result can be checked by using the volt-drop ΔV of Eq.7.21 which comes to 3.06 kV. Hence, by Eq.7.19, $E_I = 18.1$ J. This is one joule less than the anticipated value and lies well within the error band of measurements and analysis and confirms the accuracy of the ionization energy estimation technique.

We present three sets of experimental results in tables 7.3, 7.4 and 7.5, associated respectively with the three bar graphs of figures 7.24, 7.25 and 7.26. The principal variables were:

Set 1: High inductance circuit ($L = 1.24$ μH), rocket-shaped secondary projectile (R in figure 7.22), 5 cm long accelerator barrel.

Set 2: Low inductance circuit ($L = 0.31$ μH), rocket-shaped secondary projectile, 5 cm long accelerator barrel.

Set 3: Low inductance circuit (L = 0.31 µH), tumbler secondary projectile (T in figure 7.22), 2 cm long barrel.

Let the kinetic energy gain be defined by

$$Q_K = \frac{E_K}{E_A} \qquad\qquad (7.33)$$

where E_K is the kinetic energy developed in the water and measured by the momentum technique as defined by Eq. 7.31. E_A is the energy available for water acceleration, and is obtained by subtracting all known losses from the stored capacitor energy ($E_A = E_S\text{-}E_C\text{-}E_I$). In all reported experimental shots Q_K was found to be greater than one, varying between 1.03 and 3.15. This indicates a kinetic energy gain, Q_K, of the same order of magnitude as the measured pressure energy gain, Q_P, defined by Eq.7.22.

It has to be remembered, however, that the kinetic energy results are based on the assumption that all of the high velocity fog is absorbed by the balsa wood, and that the rest of the water has negligible velocity and makes no significant contribution to the acceleration of the secondary projectile. Evidence supporting the existence of slow water and its possible effects will be discussed after the presentation of the results.

Table 7.3 and figure 7.24 refer to six shots, all at V_0 =12 kV, V_t =1.8 kV and C = 0.565 µF. Three of the shots used light water (H_2O) and the other three used heavy water (D_2O). The water volume was either 1.0 or 1.5 ml.

The secondary projectile was weighed before and after the shot to determine the mass of the dry projectile, M, as well as the water mass, m, absorbed by the wood. For w = 1.0 ml, the mass m varied between 0.234 and 0.552 gm. On the whole, only about one-third of the water charge penetrated into the porous wood. This was almost certainly the fast fog component. It was followed by water travelling at lower speed which fell back on the accelerator and surrounding surfaces. Most of the residual water was found in the form of large (3-5 mm diam.) drops on the accelerator muzzle and the surroundings.

Since they were first investigated in the mid-1980s, it has been a feature of water plasma explosions that some of the water always stayed behind in the explosion cavity. This may have had something to do with the formation of plasma beads, as discussed in connection with filamentary fusion processes and particularly with figure 7.5. With a long water column, almost half the water remained in the accelerator barrel [6.9], and with shorter columns it was less. From approximate estimates to the nearest 0.1 gm, in 32 shots of 1 ml of water subjected to the kinetic energy analysis, the average amount of water remaining in the barrel was 0.2 g or 20 percent. For w =1.5 ml, this figure increased to 30 percent. It is therefore safe to assume that in all three sets of experiments, at least 20 percent of the water charge definitely did not accelerate the secondary projectile.

It is believed that high-pressure fog is being generated in a narrow channel directly above the central electrode. This was indicated by experiments carried out in the TCBOR Laboratory. A half-inch diameter Nylon (Delrin) ball was held on top of a 0.8 ml water column in the previously described accelerator. The water arc explosion inscribed a clearly defined pressure area of ¼ inch diameter on the underside of the ball and directly above the ¼ inch diameter central electrode. This proved that the lift force on the ball was mainly confined to this central region in the half-inch diameter accelerator barrel. If the electrode was made with a smaller diameter extension, a hole of the smaller diameter would be punched into the ball. Presumably it is the narrow fog column, which is generated near the central electrode, and emerges first from the muzzle and then penetrates up to 20 mm deep into the balsa wood.

Shot #	S60106	S60109	S60110	S60111	S60210	S60211
w (ml)	1.0, H_2O	1.0, H_2O	1.0, D_2O	1.0, D_2O	1.0, D_2O	1.5, H_2O
M (gm)	69.816	70.042	69.923	70.201	71.353	71.624
M+m (gm)	70.189	70.36	70.325	70.605	71.290	72.290
m (gm)	0.373	0.318	0.402	0.404	0.488	0.666
h (cm)	7.9	7.0	9.2	6.5	6.5	8.0
v_0 (m/s)	1.24	1.17	1.34	1.13	1.13	1.25
$u_{r.m.s}$ (m/s)	260.2	287.8	260.9	219.1	184.6	151.0
A (kA)	4.88	4.66	4.69	4.80	4.99	4.90
B (kA)	3.06	2.99	2.93	3.14	3.13	3.36
I_0 (kA)	5.48	5.21	5.28	5.34	5.60	5.38
T (µs)	11.6	12.2	11.5	12.7	11.7	14.3
$\int i^2 dt$ $(A^2 s)$	87.1	82.8	80.2	90.5	91.7	103.5
$R=2L/T$ (Ω)	0.214	0.203	0.216	0.195	0.212	0.173
E_S (J)	39.8	39.8	39.8	39.8	39.8	39.8
E_C (J)	18.6	16.8	17.3	17.7	19.4	18.9
E_1 (J)	16.3	18.9	18.0	18.1	15.3	18.3
E_A (J)	4.9	4.1	4.5	4.0	5.1	2.6
E_K (J)	12.6	13.2	13.7	9.7	8.3	7.6
$Q_K=E_K/E_A$	2.57	3.22	3.04	2.43	1.63	2.92

Table 7.3 : Kinetic energy results with high inductance circuit, rocket shaped projectile (R) and 5 cm barrel. $V_0 = 12$ kV, $V_c = 1.8$ kV, $C = 0.565$ µF, $L = 1.24$ µH

Sections cut from the wet balsa wood revealed that the water damage, and therefore the water velocity, varied widely from place to place. It will be assumed that the velocity distribution has the form of a half-cycle of a sine wave. The average ordinate of the sine wave is the amplitude divided by $\pi/2$, and the root-mean-square (r.m.s) value is the amplitude divided by 2, yielding

$$u_{rms} = 1.1\, u_{av} \tag{7.34}$$

Equation 7.34 has been employed in the results of tables 7.3 to 7.5, and is likely to be an underestimate. In any case, the r.m.s-factor is clearly not of great importance to the kinetic energy conclusions.

The results of table 7.3 show that an appreciable fraction of the input energy E_S is converted to heat by conduction, E_C and ionization losses, E_I. The remainder, $E_A = E_S - E_C - E_I$, is available for generating kinetic energy. In all cases $E_K > E_A$. This is the important outcome of this investigation as documented by the gain factor Q_k. The relevant quantities of energy are displayed in figure 7.24. The average kinetic energy per shot was 10.9 J, or 25 percent of the initial stored energy.

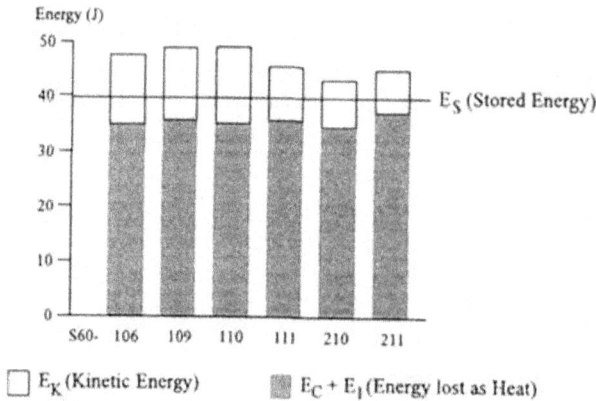

Figure 7.24 : Anomalous kinetic energy in the 6 shots described in table 7.3

On reducing the circuit inductance to 0.31 μH, the results of table 7.4 were obtained. They show an increase of the average kinetic energy to 15.6 J at 11 kV and 16.7 J at 12 kV. The kinetic energy was then approaching half of the energy supplied by the capacitors. It should be noted, however, that this was not reflected in an increased gain factor Q_K. It suggests that the advantage of the lower inductance is a reduction in energy losses in the circuit without an improvement of the conversion efficiency from available energy in the cavity to kinetic energy.

Shot #	S53006	S53007	S53010	S53011	S53012	S53013
V_0 (kV)	11	11	11	12	12	12
V_r (kV)	1.7	1.7	1.7	1.8	1.8	1.8
w (ml)	1.0, H_2O	1.0, H_2O	1.5, H_2O	1.0, H_2O	1.0, D_2O	1.5, H_2O
M (gm)	90.671	70.148	70.079	69.933	69.903	70.050
M+m (gm)	90.905	70.390	70.594	70.202	70.242	70.558
m (gm)	0.234	0.242	0.515	0.269	0.339	0.508
h (cm)	3.1	5.8	16.3	8.2	10.0	12.0
v_0 (m/s)	0.78	1.07	1.79	1.27	1.40	1.53
$u_{r.m.s}$ (m/s)	336.3	344.4	272.1	367.4	322.1	236.5
E_S (J)	33.4	33.4	33.4	39.8	39.8	39.8
E_H (J)	26	26	26	26	26	26
E_A (J)	7.4	7.4	7.4	13.8	13.8	13.8
E_K (J)	13.2	14.4	19.1	18.2	17.6	14.2
Q_K	1.78	1.95	2.58	1.32	1.28	1.03

Table 7.4 : Kinetic energy results with low inductance circuit, rocket shaped secondary projectile (R) and 5 cm barrel C = 0.565 μF, L = 0.31μH

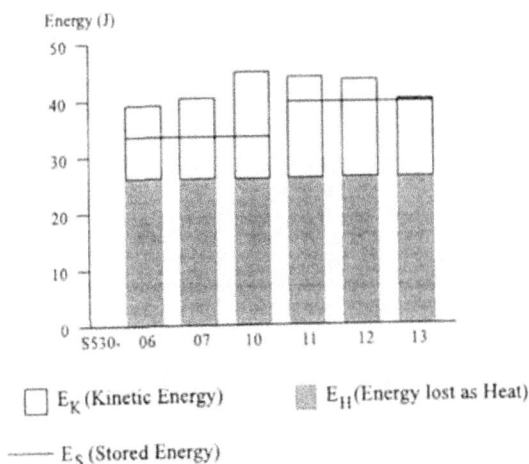

Figure 7.25 : Anomalous kinetic energy in the 6 shots described in table 7.4

Electrodynamics in the Quest for New Energy

Shot #	S60508	S60509	S60501	S60607	S60609
w (ml)	1.0, H_2O	1.0, H_2O	1.0, D_2O	1.0, H_2O	1.0, H_2O
M (gm)	51.962	52.019	52.005	47.828	50.022
M+m (gm)	52.258	52.351	52.397	48.253	50.507
m (gm)	0.296	0.332	0.392	0.425	0.485
h (cm)	39	27	32	34	32
v_0 (m/s)	2.77	2.30	2.51	2.58	2.51
$u_{r.m.s}$ (m/s)	542.1	403.9	371.7	325.5	289.6
E_S (J)	39.8	39.8	39.8	39.8	39.8
E_H (J)	26	26	26	26	26
E_A (J)	13.8	13.8	13.8	13.8	13.8
E_K (J)	43.5	27.1	27.0	22.5	20.3
Q_K	3.15	1.96	1.96	1.63	1.47

Table 7.5 : Kinetic energy results with low inductace, tumbler secondary projectile (T) and 2 cm barrel, $V_0 = 12$ kV, $V_t = 1.8$ kV, $C = 0.565$ µF, $L = 0.31$ µH

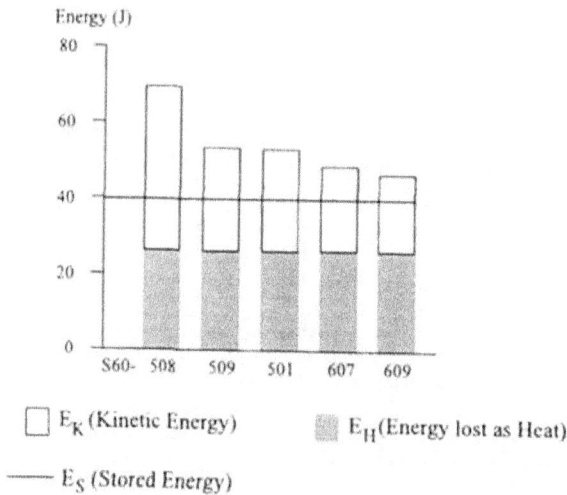

Figure 7.26 : Anomalous kinetic energy in the 5 shots described in table 7.5

The low inductance in the last two data sets, unfortunately led to severe current waveform distortion, as illustrated by the oscillogram of figure 7.27. Without being able to reliably measure the first two current peaks, a new method of estimating energy losses had to be developed. To begin with, an approximate measurement of the temperature rise in the water was made. For this purpose a piece of balsa wood was pushed 12 mm deep into the 2 cm long accelerator barrel to just above the water level. A needle thermocouple was inserted into the wood from the top. The wood was then loaded with a heavy weight so that no detectable lift could occur. Using 1 ml of distilled water, 39.8 J of energy were discharged through the circuit, as in most of the reported shots. The thermocouple indicated a temperature rise of 4.9°C. The experiment was repeated with a flying secondary projectile, and the thermocouple was inserted into the wood within five seconds of the shot. It again measured a temperature rise of approximately 5°C. The temperature was falling quite rapidly over the next 30 seconds.

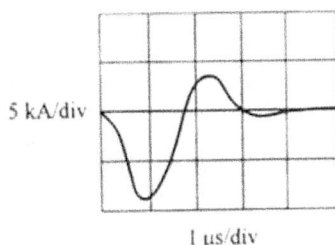

5 kA/div

1 μs/div

Figure 7.27 Severe current waveform distortion with the low inductance circuit

It is not unreasonable to assume that all of the water reached the temperature recorded by the thermocouple. This required the evolution of 5 cal = 20.9 J of heat. The precise mechanisms by which this heat was produced, whether by ionization, Joule heating, or conversion from pressure energy, is not known. None of it, however, could have contributed to the acceleration of water or the production of fog. Additional heat must have arisen in the metallic circuit and the switching arc. From prior experience it is known that the latter has to be at least five joules and could be substantially more. In analyzing the low inductance shots it will, on this basis, be assumed that 26 J of the input energy were unavailable for kinetic energy production. In tables 7.4 and 7.5, this amount of energy has been denoted by E_H. It was subtracted from E_S to obtain the energy, E_A, available for water acceleration and fog formation.

The average kinetic energy with the low inductance circuit was 16.1 J per shot, as compared with 10.9 J per shot with the high inductance circuit. Despite this increase in kinetic energy, the KE gain factor, Q_K, decreased from an average of 2.29 to 1.66 per shot in going from the higher to the lower inductance. This was probably due to an underestimate of E_H for the low inductance circuit. Without more experimental information about E_H, the tentative conclusion has to be that the lower inductance decreases the circuit energy losses, but it does not necessarily boost the conversion efficiency from available electric energy to kinetic energy.

The performance with a water charge of w = 1.5 ml was about the same as with 1.0 ml, and as in table 7.3, the heavy water showed no advantage over light water. The fact that there was a small but noticeable improvement in performance of heavy water over light water, in the pressure energy tests described earlier, seems to imply that at this moment the pressure measurements are the more accurate.

After replacing the rocket-shaped projectile R with the tumbler T of figure 7.22, and shortening the accelerator barrel from 5 to 2 cm, the greatly improved KE production indicated in table 7.5 and figure 7.26 was found. The average value of E_k per shot increased to 28 J from 16.1 J. This upgrading is believed to have been the result of eliminating the steel skirt of the R-projectile which probably interfered with its acceleration. The shorter accelerator barrel helped too, but it is thought to have made only a small contribution to the better performance. The gain factor Q_K was high, but not decidedly better than with the high inductance circuit and the more reliable loss measurements.

5.   The Effect of Slow Water on the KE Measurements

It has been pointed out that the validity of the momentum conservation formula, Eq.7.26, depends on the assumption that nothing but the absorbed water mass, m, accelerates the secondary projectile. Approximately two-thirds of the water in the explosion cavity is not absorbed by the balsa wood, and one-fifth stays behind in the accelerator. This leaves 47 percent of the water mass which flies out of the accelerator but does not penetrate into the secondary projectile. Could this slower water exert significant force on the wood?

If the KE measurements of tables 7.3 - 7.5 are correct, fog clusters travelled on the average at 300 m/s. In 100 μs, a fog cluster would have covered a distance of 3 cm. Assuming that, originally, the fog cluster was as long as the water depth in the explosion cavity, that is about one centimeter, it might well have been all absorbed in the balsa wood in less than 100 μs. With the absorbed water mass being about m = 0.33 gm, it should have resulted in an acceleration force of more than 990 N. This consideration shows that the impact force is very large, as it has to be in order to lodge the fog in the wood. Smaller forces, which are incapable of driving the water into the wood, must splash the water laterally because of the negligible shear strength of the liquid. This lack of shear strength is responsible for the inelastic collision of liquid with a solid body. The inelastic behavior can be readily observed when a jet of water strikes the flat bottom of a sink.

To investigate the behavior of slow water behind the secondary projectile, a ½-inch diameter balsa wood cylinder of 32 mm length, and weighing 0.54 gm, was pushed into an accelerator barrel of the same diameter to a depth of ½ inch. A 0.45 μF capacitor was then charged to 12 kV and discharged through 1 ml of distilled water under the balsa wood. The first 9 cm of projectile displacement was photographed with a high-speed camera at 10,000 frames per second. The displacement of the balsa wood cylinder during the first 12 frames is plotted in figure 7.28. It will be seen that after the initial acceleration, the projectile velocity was constant at 80 m/s. The initial acceleration apparently lasted for a brief period of less than 100 μs.

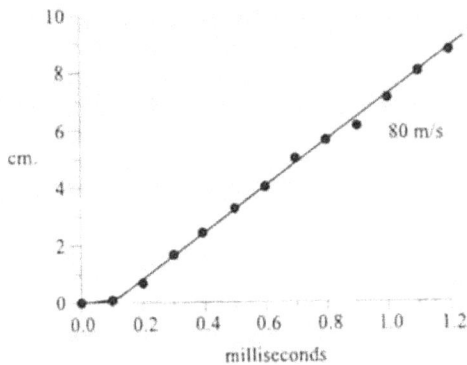

Figure 7.28 : Trajectory of 0.54 gm balsa wood cylinder fired from water arc accelerator

The high-speed photographs are difficult to reproduce for publication, nevertheless, they revealed that some water remained in contact with the wood for six centimeters of the climb. Then a gap begun to open between wood and water droplets. No sideways splashing of water occurred at any time. The droplet column was of the same diameter as the barrel and the balsa wood cylinder. The column was, however, not perfectly straight and vertical. This can be explained with the onset of projectile tumbling and some water falling from it. From the constant velocity of the secondary projectile and the appearance of the droplet column it is fair to conclude that any acceleration of the projectile by slow water was negligible.

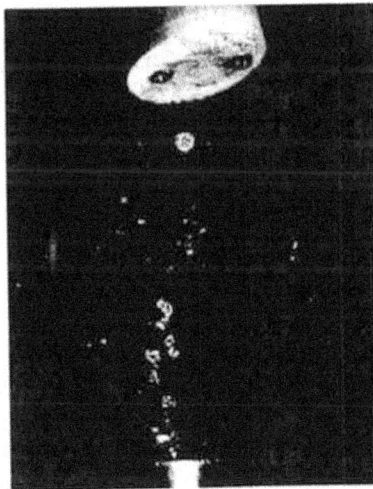

Figure 7.29 : Evidence of slow water drops below the secondary projectile

Clear evidence of the inability of the slow water to accelerate the secondary projectiles used in the kinetic energy measurements was provided by photographs such as figure 7.29. This shows the balsa wood cylinder (T in figure 7.22) of approximately 45 gm. mass about ten cm. above the accelerator muzzle. The wet area on the underside of the secondary projectile indicates where the fog has penetrated the soft wood. The relatively large water drops between muzzle and the projectile account for virtually all of the slow water which was not absorbed in the wood. The drops lag behind the projectile. They are not squashed, as they would be, had they collided with the projectile. There can be no doubt, therefore, that the reported kinetic energy results are due entirely to the fast fog, and possibly some larger mist droplets, which all entered the balsa wood and travelled with it.

6.   Summary

After more than ten years research in various laboratories it is an established fact that the forces revealed by water plasma explosions are anomalously large. This means that the forces cannot be explained by the laws of physics as they are used in 1995. As every force of nature relies on a reservoir of energy, without which the force could not exist, there must be anomalous energy released in water plasma explosions.

The same capacitor energy discharged through the same circuit does not develop anomalous forces and energy in the arc explosion when the water is replaced by atmospheric air. This is strong evidence for the anomalous energy to be internal water energy. The vacuum energy of field theory, if it were the source of the anomalous energy, would be expected to contribute in both cases. Since it does not, it is probably not a factor in water plasma explosions.

With regard to internal water energy, we may distinguish between nuclear energy, chemical bonding energy between hydrogen and oxygen atoms, and finally, the energy of liquid cohesion between H_2O molecules. The liberation of nuclear energy would almost certainly be associated with the emission of X-rays. No X-rays have been detected and, therefore, nuclear energy is unlikely to be the source of the anomalous energy.

Although no systematic search for free hydrogen and oxygen has been carried out, arcs in water are not known to have caused this chemical dissociation. After the explosion all the water can be accounted for by fog and larger drops. No rigorous proof has been provided, but all the circumstances indicate that it is not chemical energy which is responsible for the plasma explosions.

By elimination of other sources, we are left with intermolecular bonding energy as the most likely contribution to the explosions. The circumstances are sufficiently compelling to formulate the hypothesis that potential energy is liberated and that fog droplets contain less of this, per unit mass, than the water initially in the explosion cavity. The observed repulsion between fog droplets is then understandable.

A measure of the magnitude of the potential energy liberated has been obtained from measurements of the transiently stored pressure energy in the explosion cavity. It was found to be up to three times the energy supplied to the water after subtracting all the known energy losses from the capacitor energy. Subsequent kinetic energy measurements, by a method relying on momentum conservation, have indicated that a large portion of the pressure energy can be converted to kinetic energy of fog droplets. This represents the extent of our knowledge on water arc explosions at the time of writing at the end of September 1995.

7. Acknowledgements

The authors are indebted to George Hathaway, of Hathaway Consulting Services of Toronto, Canada, for helping to reopen the investigation and contributing crucial experimental results on water arc explosions. Richard Hull of the TCBOR Laboratory in Richmond, VA, demonstrated the ejection of very fast fog from water plasma accelerators. Dr. Brian Ahern and Professor Keith Johnson drew the authors' attention to the unusual properties of small clusters of matter.

Chapter 7 References

7.1   O.A. Anderson, W.R. Baker, S.A. Colgate, J. Ise, R.V. Pyle, "Neutron production in linear deuterium pinches", Physical Review, Vol.110, p.1375, 1958.

7.2   W. Lochte-Holtgreven, "Nuclear fusion in very dense plasmas obtained from electrically exploded liquid threads", Atomenergie (ATKE), Vol.28, p.150, 1976.

7.3   E. Storms, "Warming up to cold fusion", Technology Review, May/June 1994, p.19.

7.4   A.S. Bishop, *Project Sherwood*, Addison-Wesley, Reading MA, 1958.

7.5   S.K. Haendel, O. Jonsson, "Capillary fusion in very dense plasmas", Atomenergie (ATKE), Vol.36, p.170, 1980.

7.6   P. Graneau, N. Graneau, "The role of Ampère forces in nuclear fusion", Physics Letters A, Vol.165, p.1, 1992.

7.7   P. Graneau, N. Graneau, "Ampère force calculations for filament fusion experiments", Physics Letters A, Vol.174, p.421, 1993.

7.8   J.D. Sethian, A.E. Robson, K.A. Gerber, A.W. DeSilva, "Enhanced stability of neutron production in a dense z-pinch plasma from a frozen deuterium fiber", Physical Review Letters, Vol.59, p.892, 1987.

7.9   J.D. Sethian, A.E. Robson, K.A. Gerber, A.W. DeSilva, "Evolution of a deuterium fiber z-pinch driven by a long current pulse", Workshop on Alternative Magnetic Confinement Systems, Varenna, Italy, Oct. 1990.

7.10   M. Rambaut, "Capillary fusion through Coulomb barrier screening in turbulent processes generated by high intensity current pulses", Physics Letters A, Vol.163, p.335, 1992.

7.11   M.G. Haines, "Dense plasmas in z-pinches and the plasma focus", Philosophical Transactions of the Royal Society of London, Vol.A300, p.649, 1981.

7.12   G. Decker, R. Wienecke, "Plasma focus devices", Physica, Vol.82C, p.155, 1976.

7.13   H. Herold, H.J. Kaeppeler, H. Schmidt, H. Shakhatre, C.S. Wong, C. Deeney, P.Choi, "Progress in plasma focus operation up to 500 kJ bank energy", Plasma Physics and Controlled Nuclear Fusion Research, Conference Proceedings, Nice, France, Oct. 1988, International Atomic Energy Agency, Vienna, 1989.

7.14   C. Sheer, *Vistas in Science* (D.L. Arm, Editor), The University of New Mexico Press, Albuquerque NM, 1968.

7.15   I.R. Lindemuth, "Two-dimensional fiber ablation in the solid-deuterium z-pinch", Physical Review Letters, Vol.65, p.179, 1990.

7.16   E. Storms, C. Talcott-Storms, "The effect of hydriding on the physical structure of palladium and the release of contained tritium", Fusion Technology, Vol.20, p.246, 1991.

7.17   K.A. Kaliev, A.N. Baraboshkin, A.L. Samgin, E.G. Golikov, A.L. Shalyapin, V.S.Andreev, P.I.Golubnichii, "Reproducible nuclear reactions during interaction of deuterium with oxide tungsten bronze", Physics Letters A, Vol.172, p.199, 1993.

7.18   A.B. Karabut, Ya. R. Kucherov, I.B. Savvatimova, "Nuclear product ratio for glow discharge deuterium", Physics Letters A, Vol.172, p.199, 1992.

7.19   P.I. Golubnichii, V.V. Kuzminov, G.I. Merzon, B.V. Pritychenko, A.D. Filonenko, V.A. Tsarev, A.A. Tsarik, "Correlated neutron and acoustic emission from deuterium-saturated palladium target", JETP Letters, Vol.53, P.123, 1991.

7.20   P.A.M. Dirac, *The principles of quantum mechanics*, Oxford Press, London, 1947.

7.21   H.E. Puthoff, "Ground state of hydrogen as a zero-point-fluctuation determined state", Physical Review D, Vol.35, p.3266, 1987.

7.22   P. Postorino, R.H. Tromp, M.A. Ricci, A.K. Soper, G.W. Nellson, "The interatomic structure of water at supercritical temperatures", Nature, Vol.366, p.668, 1993.

7.23   R.A. Horne, *Water and aqueous solutions*, Wiley, New York, 1972.

7.24   J.B. Hasted, *Aqueous dielectrics*, Chapman and Hall, London, 1973.

7.25   N.E. Dorsey, *Properties of ordinary water substance*, Reinhold, New York, 1940.

7.26   R. Hull, "Water arc explosions", Electric Spacecraft Journal, Issue 14, 1995.

7.27   H. Kroto, "The birth of C_{60} Buckminsterfullerene" *Electronic properties of fullerenes* (H. Kuzmany et al, Editors), Springer, Berlin, 1993.

www.ingramcontent.com/pod-product-compliance
Lightning Source LLC
Chambersburg PA
CBHW061627220326
41598CB00026BA/3915